Microelectrode Arrays and Application to Medical Devices

Microelectrode Arrays and Application to Medical Devices

Editors

Colin Dalton
Alinaghi Salari

MDPI • Basel • Beijing • Wuhan • Barcelona • Belgrade • Manchester • Tokyo • Cluj • Tianjin

Editors
Colin Dalton
Electrical and Computer
Engineering Department,
University of Calgary
Canada

Alinaghi Salari
Li Ka Shing Knowledge Institute,
Ryerson University
Canada

Editorial Office
MDPI
St. Alban-Anlage 66
4052 Basel, Switzerland

This is a reprint of articles from the Special Issue published online in the open access journal *Micromachines* (ISSN 2072-666X) (available at: https://www.mdpi.com/journal/micromachines/special_issues/Microelectrode_Arrays).

For citation purposes, cite each article independently as indicated on the article page online and as indicated below:

LastName, A.A.; LastName, B.B.; LastName, C.C. Article Title. *Journal Name* **Year**, *Article Number*, Page Range.

ISBN 978-3-03943-174-8 (Hbk)
ISBN 978-3-03943-175-5 (PDF)

Cover image courtesy of Xing Yang.

Contents

About the Editors

Colin Dalton is a senior member of the IEEE and obtained his PhD from the University of Wales, UK. He specializes in micro/nano technology for biomedical applications, and is currently an assistant professor in the Electrical and Computer Engineering Department and the Biomedical Engineering Graduate Program at the University of Calgary. He is also Director of the University's Microfabrication Cleanroom facility. He is Co-Founder and CTO of Neuraura Biotech Inc., a spin-off company developing brain machine interfaces. He previously worked for the UK Lab-On-A-Chip Consortium, a collaboration between industry, government and academia, which developed new microsystems from university research. He is the author of over 80 journal and conference proceedings, and has four patents pending. His current research interests include electrokinetics, microneedles, lab-on-a-chip systems, nerve regeneration through electrical stimulation, and brain machine interfaces. His research was featured as a Canada-wide success story by the federal not-for-profit organization, CMC Microsystems.

Alinaghi Salari obtained his MSc degrees from the University of Calgary, Canada, and Sharif University of Technology, Iran. He obtained his PhD from Ryerson University, Canada, where he worked on acoustofluidics for biomedical applications. His research interests include microfluidic techniques and lab-on-a-chip strategies.

micromachines

Editorial

Editorial on the Special Issue on Microelectrode Arrays and Application to Medical Devices

Alinaghi Salari [1] and Colin Dalton [2],*

[1] Institute for Biomedical Engineering, Science and Technology (iBEST), Toronto, ON M5B 1T8, Canada; a1salari@ryerson.ca
[2] Electrical and Computer Engineering Department, University of Calgary, Calgary, AB T2N 1N4, Canada
* Correspondence: cdalton@ucalgary.ca; Tel.: +1-403-210-8464

Received: 11 August 2020; Accepted: 13 August 2020; Published: 14 August 2020

In this editorial note, we briefly review the major findings of the 10 articles published in the Special Issue on microelectrode arrays and application to medical devices. The articles are categorized into three sections, i.e., fabrication techniques [1–4], application demonstration [5–8], and review of recent advances in this field [9,10].

1. Fabrication Techniques

In many drug screening and electrophysiological applications, microelectrodes are commonly used to study electric field and impedimetric measurements of cells in vitro. In order to couple such a platform with optical microscopy, indium tin oxide (ITO), which is a transparent material, can be used instead of titanium nitride (TiN). The deposition of ITO and TiN thin films are commonly conducted by the sputter deposition technique. However, Ryynänen et al. successfully demonstrated that ion beam-assisted electron deposition (IBAD) can be used as an alternative method for this purpose [1]. They studied three different approaches for the fabrication of microelectrodes. First, an optimal imaging capability was demonstrated when the tracks and electrodes were both made of ITO. In the second approach, a thin layer of TiN was coated onto the electrodes in order to decrease the electrical impedance. To obtain the optimal electrical performance, in the third approach, the electrodes were made of opaque TiN electrodes. In their study, the electrical impedance and noise levels, light transmission through the substrate, and biocompatibility of the fabricated devices were also analyzed.

Laser micromachining is another technique used for electrode microfabrication, which involves the ablation of materials, and are available in a wide range of wavelengths, pulse durations, and repetition rates. When operated in the femtosecond regime, this technique can be useful for the fabrication of sub-10-micron features. However, this operational regime can be costly and suffers from high power consumption. Hart et al. characterized a multimodal laser micromachining tool featuring limited power consumption and nanosecond ablation [2]. They showed that these features provide better control over the material selectivity for laser micromachining and ablation depths. They also analyzed the ablation characteristics of six different materials and demonstrated the application of this tool for the laser micromachining of shadow masks with a 1.5-μm feature size and microelectrodes with 7-μm electrode gas widths.

For the application of implantable microelectrode arrays, the material and biocompatibility of such devices are crucial. Intracortical microelectrode arrays, for example, can be used for recording electrical function or neural stimulation in brain tissue. Due to the disadvantages of the application of stiff and brittle materials for the fabrication of such implants, polymer materials, which enable robust, yet softer, properties have been suggested. Still, a successful implant of a polymer-based device in brain tissue can be challenging and, as a result, further fabrication steps including temporary stiffening coatings or insertion guides are often required. To overcome these limitations, Stiller et al. have investigated the use of shape memory polymers (SMP) in intracortical devices [3]. They fabricated

an intracortical recording probe using a thiol-ene/acrylate formulation, which features a dynamic response to stimuli from the environment, meaning that the modulus of elasticity of such material changes considerably (by two orders of magnitude) when its environment is changed from room temperature and dry conditions to body temperature and wet conditions. The authors also conducted performance characterization and demonstrated in vivo intracortical recordings over several weeks in a rat brain, suggesting that this SMP device is a reliable candidate for such applications.

In another study, Wang et al. investigated the fabrication of a microelectrode array in the form of pyramidal microstructures on a polydimethylsiloxane (PDMS) substrate with a parylene transition layer between the electrode and the substrate [4]. Their results showed that this method of fabrication can increase the contact area of the electrodes, enhance the bonding force between the substrate and the electrodes, and reduce the electrode impedance as well as the electrode–skin contact impedance. This fabrication method can help the development of wearable and implantable medical devices.

2. Application Demonstration

Monitoring the local cerebral tissue oxygen levels ($PbtO_2$) can provide critical information about brain function including neurovascular metabolic activity. Different invasive and non-invasive techniques have been reported for measuring the tissue oxygen levels. Among those, the amperometric sensing technique allows real-time monitoring of brain oxygenation and can be an effective bedside method for the detection of cerebral ischemia. Due to the electrocatalytic behavior of Pt, Ledo et al. reported the potential application of those commercially available intracranial recording electrodes that include a thin Pt film on their recording sites, for amperometric monitoring of $PbtO_2$ [5]. They conducted a surface morphology characterization and an electrochemical evaluation of the performance of such a device for oxygen detection and concluded that these probes can be repurposed for in vivo multisite monitoring of $PbtO_2$.

The study of cellular electrophysiology in in vitro settings can be performed using microelectrode arrays, which can conduct extracellular recordings of various chemicals, such as glucose. This application was demonstrated by Alassaf et al., who studied the function of human islets by measuring the electrical activities of dissociated islet cells in real-time and over prolonged culture periods using microelectrode arrays [6]. Compared with glucose-stimulated insulin secretion assays, their results showed a correlation between the electrical activities and the functional secretory response of the islet cells. Their measurements using a microelectrode array platform exhibited a highly sensitive and rapid assessment of islet functionality, demonstrating its potential application for studying islet physiology.

Another important application of microelectrodes is in biosensing technology. Bisphenol A (BPA) is an organic chemical which can be detrimental to various organ systems including the endocrine, reproductive, and nervous systems. In an effort to enhance the detection capabilities of BPA in serum samples, Luo et al. coupled self-assembly technology with AC electrokinetics (ACEK) and showed that this strategy could yield a rapid and sensitive detection of BPA antigens in fluid samples [7]. In their work, the surfaces of microelectrodes were functionalized with the BPA antibody via self-assembly, while an AC electric field was used to actuate the electrodes. Compared to diffusion-limited sensing approaches, the ACEK microflows enhanced the transport of antigens and, thus, improved the antigen-antibody binding. The interfacial capacitance was then measured in order to analyze the concentration of the bound antigens. This process resulted in the successful detection of nanomolar levels of BPA within one minute of operation, signifying its potential application in the detection of biochemical molecules.

In another study on electrokinetics, the transient motion of Janus particles acting as mobile microelectrodes was studied by Liu et al. [8]. A Janus particle is a type of particle that has a nonuniform polarizable body, such that half of its body is less polarizable than the surrounding liquid, while the other half is more polarizable. With this special configuration, particle translation due to nonlinear induced-charge electrophoresis (ICEP) can occur even in the presence of a uniform electric field.

In this work, the authors numerically investigated the motion of cylindrical Janus polystyrene particles with half bodies covered with thin films of a polarizable metal (e.g., gold). In their 2D simulation, the transient electrokinetic behavior of these particles was studied by considering the induced-charge electroosmotic flow and Maxwell–Wager interfacial polarization, assuming that the particles were freely suspended in an electrolyte far from the channel walls. They reported a new electrokinetic transport phenomenon, called ego-dielectrophoresis, which is dominant at high frequency ranges (e.g., 100 MHz). They also showed that this new electrokinetic phenomenon causes the direction of particle motion to reverse as the actuation frequency transitions from low (e.g., 100 Hz) to high (e.g., 100 MHz) ranges. This AC electrokinetics of Janus microelectrodes can be particularly useful for the detection of biomolecules in lab-on-a-chip devices.

3. Review of Recent Advances in the Field

As an emerging technology, wearable and flexible sensing have been widely studied due to their important role in personalized medicine and their capabilities in real-time monitoring of an individual's health. The application of liquid metals, which can provide a higher degree of flexibility and conductivity compared to conventional soft sensors, in different biomedical areas was reviewed by Ren et al. [9]. The authors presented the recent progress of liquid metal-enabled soft sensors from different aspects, including innovations in material selection and fabrication techniques, fundamental principles, and typical applications. The existing challenges and possible development directions in this area were also pointed out in their article.

As discussed earlier, the AC electrokinetic effect can be useful for sample enrichment and fluid and particle transport in microfluidic devices. In a review article, Salari et al. discussed recent advances in this field with a focus on AC electrothermal techniques from different aspects involving external electric field configurations, temperature field, and the resultant velocity field [10]. The equations governing electrothermal effect were first presented, and then various strategies for the generation of electrothermal microflows were reviewed. The major applications of such flows in microfluidics and key points in numerical simulations and experimental AC electrothermal devices were covered. The authors also discussed some current limitations and future directions in this research field.

At the end of this editorial note, we would like to thank all the authors who contributed to this Special Issue and also to acknowledge the time and effort dedicated by the reviewers to improve the quality of the published articles.

Funding: This research received no external funding.

Conflicts of Interest: The authors declare no conflict of interest.

References

1. Ryynanen, T.; Mzezewa, R.; Meriläinen, E.; Hyvärinen, T.; Lekkala, J.; Narkilahti, S.; Kallio, P. Transparent microelectrode arrays fabricated by ion beam assisted deposition for neuronal cell in vitro recordings. *Micromachines* **2020**, *11*, 497. [CrossRef] [PubMed]
2. Hart, C.; Rajaraman, S. Low-power, multimodal laser micromachining of materials for applications in sub-5 μm shadow masks and sub-10 μm interdigitated electrodes (IDEs) fabrication. *Micromachines* **2020**, *11*, 178. [CrossRef] [PubMed]
3. Stiller, A.M.; Usoro, J.O.; Lawson, J.; Araya, B.; González-González, M.A.; Danda, V.R.; Voit, W.E.; Black, B.J.; Pancrazio, J.J. Mechanically robust, softening shape memory polymer probes for intracortical recording. *Micromachines* **2020**, *11*, 619. [CrossRef] [PubMed]
4. Wang, S.; Yan, J.; Zhu, C.; Yao, J.; Liu, Q.; Yang, X. A low contact impedance medical flexible electrode based on a pyramid array micro-structure. *Micromachines* **2020**, *11*, 57. [CrossRef] [PubMed]
5. Ledo, A.; Fernandes, E.; Quintero, J.E.; Gerhardt, G.A.; Barbosa, R.M. Electrochemical evaluation of a multi-site clinical depth recording electrode for monitoring cerebral tissue oxygen. *Micromachines* **2020**, *11*, 632. [CrossRef] [PubMed]

6. Alassaf, A.; Ishahak, M.; Bowles, A.; Agarwal, A. Microelectrode array based functional testing of pancreatic islet cells. *Micromachines* **2020**, *11*, 507. [CrossRef] [PubMed]
7. Luo, H.; Lin, X.; Peng, Z.; Song, M.; Jin, L. Rapid and sensitive detection of bisphenol a based on self-assembly. *Micromachines* **2020**, *11*, 41. [CrossRef] [PubMed]
8. Liu, W.; Ren, Y.; Tao, Y.; Yan, H.; Xiao, C.; Wu, Q. Buoyancy-free janus microcylinders as mobile microelectrode arrays for continuous microfluidic biomolecule collection within a wide frequency range: A numerical simulation study. *Micromachines* **2020**, *11*, 289. [CrossRef] [PubMed]
9. Ren, Y.; Sun, X.; Liu, J. Advances in liquid metal-enabled flexible and wearable sensors. *Micromachines* **2020**, *11*, 200. [CrossRef] [PubMed]
10. Salari, A.; Navi, M.; Lijnse, T.; Dalton, C. AC electrothermal effect in microfluidics: A review. *Micromachines* **2019**, *10*, 762. [CrossRef] [PubMed]

 *micromachines*

Article

Electrochemical Evaluation of a Multi-Site Clinical Depth Recording Electrode for Monitoring Cerebral Tissue Oxygen

Ana Ledo [1,2,3,*], Eliana Fernandes [1], Jorge E. Quintero [4], Greg A. Gerhardt [4,5] and Rui M. Barbosa [1,2,3]

1   Faculty of Pharmacy, University of Coimbra, 3000-548 Coimbra, Portugal; eliana.fernandes2604@outlook.com (E.F.); rbarbosa@ff.uc.pt (R.M.B.)
2   Center for Neuroscience and Cell Biology, University of Coimbra, 3004-504 Coimbra, Portugal
3   Center for Innovative Biomedicine and Biotechnology, University of Coimbra, 3004-504 Coimbra, Portugal
4   Department of Neurosurgery, University of Kentucky Medical Center, Lexington, KY 40536, USA; george.quintero@uky.edu (J.E.Q.); gregg@uky.edu (G.A.G.)
5   Department of Neuroscience, University of Kentucky Medical Center, Lexington, KY 40536, USA
*   Correspondence: analedo@cnc.uc.pt

Received: 3 June 2020; Accepted: 25 June 2020; Published: 28 June 2020

Abstract: The intracranial measurement of local cerebral tissue oxygen levels—$PbtO_2$—has become a useful tool for the critical care unit to investigate severe trauma and ischemia injury in patients. Our preliminary work in animal models supports the hypothesis that multi-site depth electrode recording of $PbtO_2$ may give surgeons and critical care providers needed information about brain viability and the capacity for better recovery. Here, we present a surface morphology characterization and an electrochemical evaluation of the analytical properties toward oxygen detection of an FDA-approved, commercially available, clinical grade depth recording electrode comprising 12 Pt recording contacts. We found that the surface of the recording sites is composed of a thin film of smooth Pt and that the electrochemical behavior evaluated by cyclic voltammetry in acidic and neutral electrolyte is typical of polycrystalline Pt surface. The smoothness of the Pt surface was further corroborated by determination of the electrochemical active surface, confirming a roughness factor of 0.9. At an optimal working potential of −0.6 V vs. Ag/AgCl, the sensor displayed suitable values of sensitivity and limit of detection for in vivo $PbtO_2$ measurements. Based on the reported catalytical properties of Pt toward the electroreduction reaction of O_2, we propose that these probes could be repurposed for multisite monitoring of $PbtO_2$ in vivo in the human brain.

Keywords: brain tissue oxygen; in vivo monitoring; multi-site clinical depth electrode

1. Introduction

Monitoring local cerebral tissue oxygen levels—$PbtO_2$—is increasingly used in neurological intensive care units to guide therapeutic strategies aimed at maintaining O_2 levels above threshold, namely in patients suffering of severe acute brain conditions such as traumatic brain injury (TBI) and aneurysmal subarachnoid hemorrhage (aSAH), and during certain neurosurgery procedures [1–3]. The rationale for this comes from studies showing improved outcome in patients with brain O_2 monitoring and $PbtO_2$ targeted therapeutic approaches [4,5]. Monitoring of $PbtO_2$ has allowed the determination of the normal range of $PbtO_2$ in the healthy brain tissue to be 25–30 mmHg [6–9], while values below 15 mmHg are typically associated with hypoxia and ischemia [10,11].

Due to the dependency of neuronal activity over oxidative metabolism for energy supply, monitoring $PbtO_2$ can be a surrogate signal of neurovascular response and metabolic activity, both of which can become severely compromised in situations of acute brain injury [12,13]. In addition, cerebral ischemia is linked to the onset of secondary brain injury by predisposing brain tissue to an energetic crisis as well as contributing to the initiation of spreading cortical depolarizations [14,15]. These are all-or-none tissue level events characterized as near-complete breakdown of neuronal membrane potential which, in injured tissue, can initiate cell death cascades [16]. Finally, besides being a key metabolite in energy metabolism, O_2 is also a substrate for multiple enzymes within cells and can act as a signal for genetic adaptation to situations of hypoxia, regulating gene expression via hypoxia-inducible factor (HIF) dependent pathways [17].

Currently approved methods for clinical monitoring of $PbtO_2$ include invasive techniques such as amperometric sensors and optical sensors, and non-invasive techniques such as those using near infrared spectroscopy (NIRS) [18]. NIRS does not directly measure O_2, but rather the level of hemoglobin saturation within a given tissue volume [19]. Among the invasive techniques, the amperometric LICOX® probe by Integra® LifeScience is considered by some to be the gold standard for $PbtO_2$ monitoring [20]. This is a Clark-type sensor comprising cathode and anode electrodes encased in an 80 μm-thick polyethylene membrane [21] across which tissue O_2 diffuses to then be detected in an electrochemical reduction reaction, producing an analytical redox current signal. Optical probes such as Neurovent®-PTO by Raumedics® are based on fluorescence quenching by O_2 of a probe [22]. Each one of these approaches has advantages and limitations, as expected with any other sensing technology. The amperometric and optical probes for focal $PbtO_2$ and NIRS probes allow real-time monitoring of brain oxygenation in patients with variable temporal and spatial resolution. Both focal probes are invasive and are single site recording devices, and thus positioning of the probe limits the information obtained. On the other hand, NIRS probes are non-invasive and can assess several regions simultaneously but can experience contamination of the signal due to extracerebral circulation. Finally, focal $PbtO_2$ probes are considered the most effective bedside method for detection of cerebral ischemia, which is not true for NIRS probes due to lack of standardization between commercial devices and undetermined threshold for ischemia [23].

Platinum recording surfaces display excellent properties for both stimulation and recording electrodes [24]. Clinical grade intracranial recording electrodes used for invasive monitoring of brain tissue electrical activity are typically composed of Pt on the recording sites [25]. This includes strip and grid subdural multi-electrode devices as well as intracranial multi-site recording electrodes for depth recordings. Considering that Pt displays electrocatalytic behavior toward the electroreduction of O_2 [26,27], we propose that clinical grade recording electrodes can be used for amperometric monitoring of $PbtO_2$, an application that has not, to the best of our knowledge, been explored.

In the current work, we have characterized the electrochemical properties of an Auragen^{TM} depth electrode (Integra® LifeScience, Princeton, NJ, USA) approved for brain mapping comprised of 12 cylindrical Pt recording sites. Furthermore, we investigated the analytical performance properties toward the reduction of O_2. This is a critical first step towards the scaling up of our previous work in rodent models aimed at establishing fast sampling amperometry coupled to multi-site electrodes as a tool for concurrent electrophysiology and electrochemistry in clinical settings such as the neurocritical care unit and neurosurgery.

2. Materials and Methods

Reagents and Solutions: All reagents used were analytical grade and obtained from Merck (Algés, Portugal). Unless otherwise stated, all in vitro electrode evaluations were performed in PBS Lite solution, 0.05 M, pH 7.4 with the following composition: 10 mM Na_2HPO_4, 40 mM NaH_2PO_4, and 100 mM NaCl. Saturated O_2 solutions for electrode calibration were prepared by bubbling PBS with 100% O_2 (Air Liquide, Algés, Portugal) for 20 min, resulting in an O_2 solution of 1.3 mM concentration at 22 °C [28]. Removal of O_2 from solutions was achieved by purging with N_2 (Air Liquide) for at least 20 min.

Auragen™ Depth Electrode: In the current study we used a clinical grade flexible Auragen™ depth electrode (ref. AU12D5L25) comprising 12 cylindrical Pt recording contacts with 2.5 mm in length and 5 mm spacing between consecutive recording sites (Figure S1) and gold connector contacts. The geometrical area of each lead was calculated to be 0.094 cm^2 based on the measured diameter of 1.2 mm. The probe was used with no surface modifications or treatments.

Scanning Electron Microscopy and Elemental Composition Analysis: High-resolution scanning electron microscopy (SEM) was performed using a field emission scanning electron microscope coupled with energy dispersive X-ray spectroscopy (EDS) (Zeiss Merlin coupled to a GEMINI II column). The elemental composition was obtained from backscattered electron detection using EDS at 10 keV (X-Max, Oxford Instruments, High Wycombe, UK). Conductive carbon adhesive tabs were used to ground the electrode surface and secure the sample onto the specimen holder.

Electrochemical Instrumentation: Electrochemical characterization was performed on a MultiPalmSens4 Potentiostat equipped with a MUX8-R2 Multiplexer (PalmSens BV, Houten, The Netherlands) and controlled by MultiTrace software (PalmSens BV, The Netherlands). We used a three-electrode electrochemical cell comprising the depth electrode as working electrode, Ag/AgCl in 3 M NaCl as reference electrode (RE-5B, BAS Inc, West Lafayette, IN, USA) and a Pt wire as auxiliary electrode.

Electrode Calibration: The depth electrodes were calibrated to assess analytical performance toward O_2 response. Calibrations were performed in 0.05 M PBS Lite pH 7.4 (20 mL) at room temperature (22 °C) with continuous stirring at low speed (240 rpm). Oxygen was removed by purging the solution with N_2 for a minimum period of 20 min, after which the needle was removed from the solution and kept above the surface to decrease O_2 back-diffusion. Once a stable baseline was obtained, 8.25 µM aliquots of the O_2 saturated solution were added in 7 consecutive repetitions (concentration range 0–57.75 µM). The mean recording display frequency was set at 4 Hz.

Data Analyses: Data analyses were performed using MultiTrace (PalmSens BV, Houten, The Netherlands), OriginPro 2016 (OriginLab, Northampton, MA, USA) and GraphPad 5.0 (GraphPad Software, San Diego, CA, USA). Values are given as the mean ± SD. The number of repetitions is indicated in each individual determination. The sensitivity of depth electrode sites towards O_2 reduction was determined by linear regression analysis in the range 0–60 µM. The limit of detection (LOD) was defined as the concentration that corresponds to a signal-to-noise ratio of 3, calculated using the expression:

$$LOD = 3 \times SD/m,$$

where SD is the standard deviation of the baseline (20 s interval) and m is the slope of the calibration curve obtained [29].

3. Results

3.1. Characterization of the Electrode Surface—Morphology and Chemical Analysis

To evaluate the morphology of the Pt surface of the depth electrode, we obtained SEM micrographs of recording sites. As shown in Figure 1A and B, the Pt surface or the recording site appears to be smooth. Higher amplification revealed that the Pt coverage is not completely uniform throughout the surface (Figure 1C). The elemental composition of the surface of the recording sites was analyzed by energy dispersive X-ray spectroscopy (EDS). As shown in Figure 1D, the active surface is primarily composed of Pt (approx. 80%), although C, O, and Al were also found to be present in lower proportions (approx. 9, 7, and 5%, respectively). To further investigate the composition of the surface, we determined the elemental composition of different regions of the surface, as indicated in Figure 1E. The lighter regions of the SEM (labeled "Spectrum 10" in Figure 1E) are composed primarily of Pt (approx. 93%) with some C contamination (Figure 1F), while the darker regions (labeled "Spectrum 9" in Figure 1E) are composed of Al and O (50 and 44%, respectively, Figure 1G). The pseudo-color map of the relative distribution of different elements on the EDS-analyzed surface (Figure 2) revealed that C is uniformly distributed over the surface and is likely a contaminant. Furthermore, Pt distribution is complimentary of that of Al and O, which overlap, suggesting that the Pt has been deposited over an aluminum oxide surface (most likely Al_2O_3), although small regions are not coated with Pt. Comparison of the surface morphology and elemental composition before and after electrochemical evaluation revealed no significant differences, indicating a stable Pt surface (Supplementary Figure S2).

3.2. Electrochemical Active Surface Area

A standard electrochemical redox couple was used to determine the electrochemical behavior of the Pt surface of the recording sites of the depth electrode. Cyclic voltammetry was carried out in 5.0 mM hexaamineruthenium (III) chloride ($Ru(III)(NH_3)_6$) in 0.5 M KCl solution at scan rates from 25 to 200 mV s^{-1}. As shown in Figure 2, the cyclic voltammograms revealed a conventional cyclic voltammetry behavior with well-defined symmetrical oxidation and reduction peaks appearing at 25 mV s^{-1}. In addition, both the anodic and cathodic peak currents ($I_{p,a}$ and $I_{p,c}$, respectively) varied linearly with the square root of the scan rate (Figure 3 inset; R^2 values of 0.999 for both $I_{p,a}$ and $I_{p,c}$) indicating that the process was diffusion-controlled. The average $I_{p,a}/I_{p,c}$ ratio was 0.8 ± 0.2 (n = 12), which is close to the theoretical value of 1 for a totally reversible reaction [30]. The $E_{1/2}$ and $E_{pa} - E_{pc}$ values were determined to be 182 ± 3 mV and 72 ± 2 mV (n = 12), respectively.

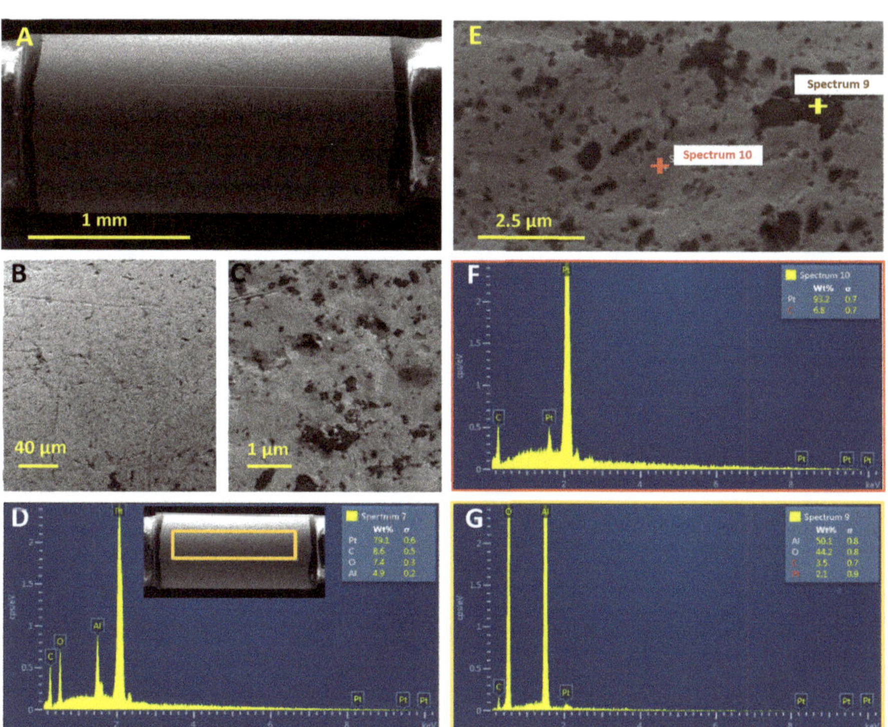

Figure 1. (**A**) General view of a recording site on the Auragen^{TM} Probe; (**B**) and (**C**) High-resolution micrographs of the recording surface; (**D**) EDS elemental analysis of the surface (ROI indicated in inset with orange rectangle); (**E**) Different regions of the recording surface were targeted for EDS elemental analysis, indicated with red and yellow "+" signs, respectively; (**F**) and (**G**) show the elemental composition spectrum of the lighter (conductive) regions (predominantly Pt) and of the darker (non-conductive) region (predominantly Al and O), respectively.

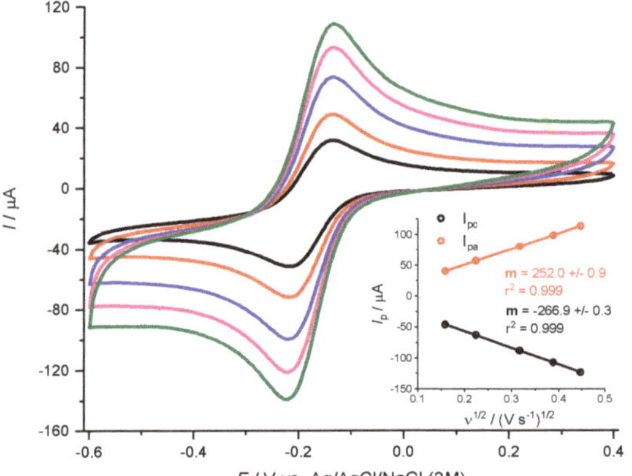

Figure 2. Pseudo-color map of the relative distribution of different elements on the EDS analyzed surface. Blue—platinum; green—aluminum; red—carbon, yellow—oxygen. Pt and Al distribution are complementary. Note that Al and O overlap, suggesting Al-O as a substrate for the Pt overcoat. Uniform distribution of C is in line with contamination.

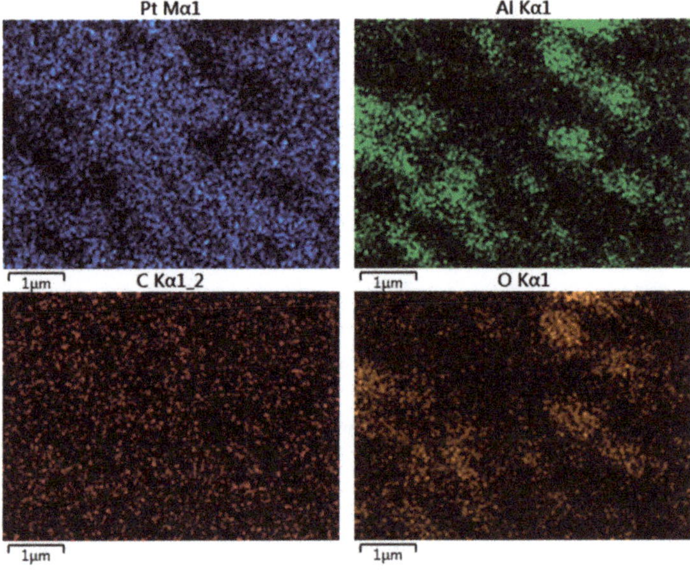

Figure 3. Reversible redox reaction of Ru(III)(NH$_3$)$_6$ in 0.5 M KCl at increasing scan rates (from 20 mV s^{-1} in black to 200 mV s^{-1} in green) and respective Ip vs. $v^{1/2}$ plot for determination of the electrochemical surface area of the depth electrode recording site.

The electrochemically active surface area of the Pt recording sites was estimated using the Randles–Sevick equation for a reversible oxidation-reduction reaction considering a diffusion coefficient of $D = 7.1 \times 10^{-6}$ cm^2 s^{-1} [31]. The calculated surface area was $8.5 \times 10^{-2} \pm 1.0 \times 10^{-2}$ cm^2 corresponding to a surface roughness of 0.90 ± 0.1 ($n = 12$). This is in line with the smooth Pt surface observed in the SEM micrographs, as well as the Pt coverage determined from the EDS elemental analysis, which showed that the Pt coverage of the recording site is roughly 90% of the analyzed area, with Al and O making up most of the remaining area.

3.3. Electrochemical Behavior in Acidic Electrolyte and in Neutral PBS

The well-known characteristic cyclic voltammogram of Pt in acid solution was used to further examine the electrochemical behavior of the Pt recording site of the depth electrode. For this purpose, the probe was characterized by cyclic voltammetry in N_2-purged H_2SO_4 (0.1 M). Figure 4A shows cyclic voltammograms recorded between −0.4 and 1.4 V *vs* Ag/AgCl at increasing scan rates (50–1000 mV s^{-1}) of a single recording site. The typical cyclic voltammogram exhibited redox peaks at −0.08 and −0.2 V. Furthermore, the presence of the three distinct peaks for H^+ desorption was clearly observed. An oxidation wave was observed for $E > 0.5$ V due to the formation of Pt oxide species Pt-O and Pt-OH, and there is a strong reduction peak at about 0.52 V corresponding to oxide reduction.

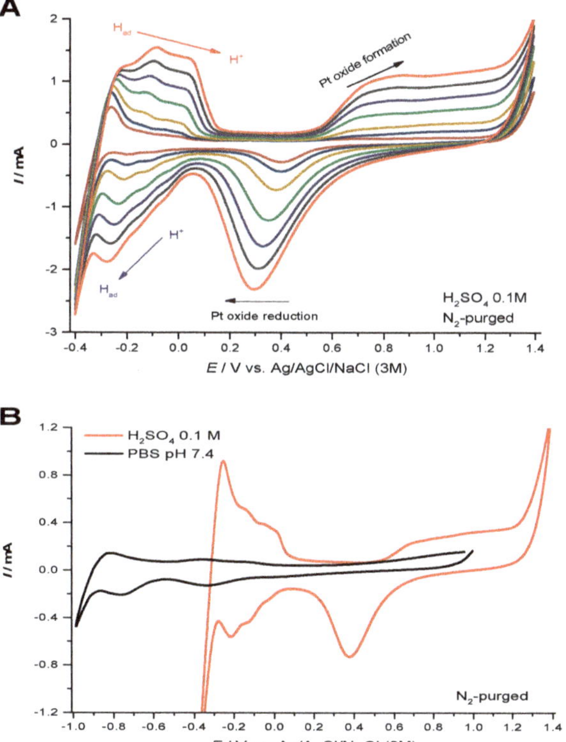

Figure 4. Electrochemical behavior in acidic and neutral electrolyte media. (**A**) Successive cyclic voltammograms (25th scan) at increasing scan rates (50–1000 mV s^{-1}) obtained in N_2 saturated 0.1 M H_2SO_4, detailing the typical Pt oxide formation and reduction, proton adsorption (2 peaks) and desorption (3 peaks), and double layer zones. (**B**) Comparative CV plots (0.2 V s^{-1}) recorded in N_2-saturated 0.05 M, pH 7.4 PBS (black line) and N_2-saturated 0.1 M, H_2SO_4 (red line) highlighting the positive shift in hydrogen evolution potential and increasing currents for Pt-oxide formation and reduction at lower pH on the Pt surface of the Integra Probe recording site.

We further characterized the electrode behavior in a neutral physiological-like media (0.05 M PBS Lite at pH 7.4), which simulates brain extracellular fluid. As shown in Figure 4B, increasing the electrolyte pH resulted in the expected negative shift in hydrogen adsorption/desorption and Pt-O formation/reduction peaks as well as a decrease in peak current width. In both electrolytes, the potential window—that is, the potential range between molecular hydrogen evolution and the evolution of molecular oxygen—is approximately 1.5 V.

3.4. Electrochemical Impedance Spectroscopy

Electrochemical impedance spectroscopy (EIS) allows the study of the physical and interfacial properties of electrochemical systems. Spectra were recorded in a N_2-purged solution containing 5 mM $K_4[Fe(II)(CN)_6]$ and 5 mM $K_3[Fe(III)(CN)_6]$ in KCl 0.5 M at room temperature by applying a sinusoidal wave of amplitude 10 mV between 100 kHz and 0.1 Hz (10 frequencies per decade) at the OCP (+0.24 V vs. Ag/AgCl). Before recording each spectrum, the electrode was held at this applied potential for 5 minutes.

The Bode plot (Figure 5A) and complex plane plot (Figure 5B) display the expected profile for a single step charge transfer process with diffusion of the reactants to the electrode surface. The complex plane plot shows the typical capacitive arc at high frequencies followed by a straight line (45°) at lower frequencies. The data were fitted to the Randles circuit [32] shown in the inset of Figure 5B, and consisting of the cell resistance (R_1) in series with a combination of a constant phase element (Q_1) in parallel with the series combination of a charge transfer resistance (R_2) and a Warburg impedance element (W). The latter accounts for mass transfer limitations imposed by diffusion, which appear at lower frequencies. The values for the charge transfer resistance, Warburg coefficient and double layer capacitance from fitting to the equivalent electrical circuit are presented in Table 1.

Table 1. Summary of fitted parameter results for impedance spectroscopy measurements ($N = 10$) [a].

R1 (Ω)	R2	Q	n	AW ($\Omega \cdot s^{-0.5}$)	Z at 1 kHz
18.98 ± 2.06	$133.5 \pm 33.4 \ \Omega$	$47.39 \pm 37.99 \ \mu F \ s^{n-1}$	0.68 ± 0.05	257 ± 14	$77.44 \pm 20.55 \ \Omega$
	$9.3 \pm 2.2 * \Omega \ cm^2$	$0.65 \pm 0.49 * mF \ s^{n-1} \ cm^{-2}$			$5.4 \pm 1.3 * \Omega \ cm^2$

[a] Values shown with * are normalized by calculated surface area.

The value of Z' at 1 kHz is typically reported for impedance measurements on electrodes. In this work, the recording sites of the depth electrode showed a Z' value of $5.4 \pm 1.3 \ \Omega \cdot cm^2$ ($77.4 \pm 20.6 \ \Omega$ before area normalization) at 1 kHz.

3.5. Oxygen Reduction Reaction at the Platinum Surface

To determine the most suitable working potential for monitoring O_2 in vivo, we performed calibrations of the depth electrode at applied potentials ranging from 0.0 to −0.8 V vs. Ag/AgCl. The slopes corrected for electrochemical surface area are plotted in Figure 6A, revealing an increase in sensitivity as the applied potential decreased from 0.0 to −0.8 V. Although it seems tempting to use the highest value of −0.8 V, as seen in Figure 6B, the increase in sensitivity is accompanied by an increase in the baseline current. Considering the expected low values of brain tissue O_2 and variations in the μM range, to optimize resolution and LOD, we chose −0.6 V vs. Ag/AgCl as the optimal applied potential. In Figure 6C, we show a representative calibration recording and respective calibration curve (inset). We observed linearity for the concentration range 0–50 μM and mean sensitivity of $−1.2 \pm 0.2$ A $M^{-1} \cdot cm^{-2}$ and mean LOD of 0.4 ± 0.1 μM ($n = 4$).

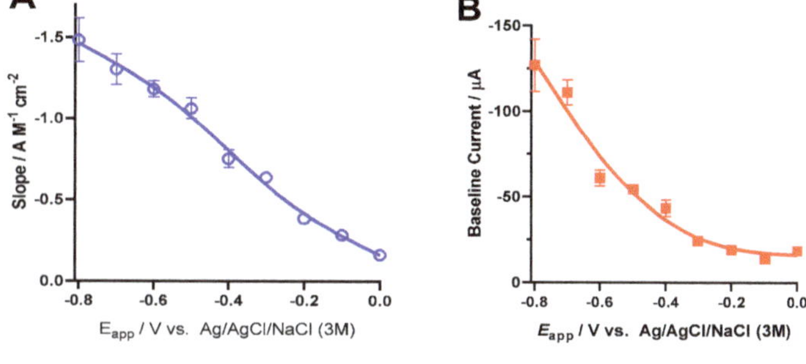

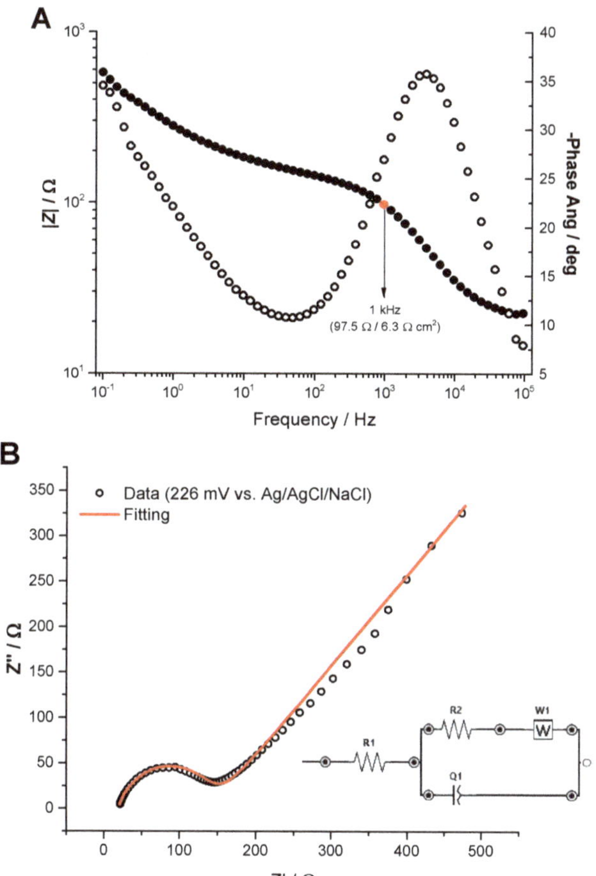

Figure 5. Electrochemical impedance spectroscopy measurements obtained in $K_4[Fe(II)(CN)_6]/K_3[Fe(III)(CN)_6]$ (5 mM) in KCl 0.5 M. (**A**) Impedance–frequency plot (Bode plot). Filled circles represent $|Z|$ values, and open circles are those obtained for the phase shift. The red circle highlights the $|Z|$ value at 1 kHz. (**B**) Complex plane electrochemical impedance spectrum (Nyquist plot) of experimental data (open circles). Red line shows fitting to the electrical equivalent circuit shown in the inset. R1, solution resistance; R2, electron or charge transfer resistance; W, Warburg impedance element; Q, constant phase element.

Figure 6. *Cont.*

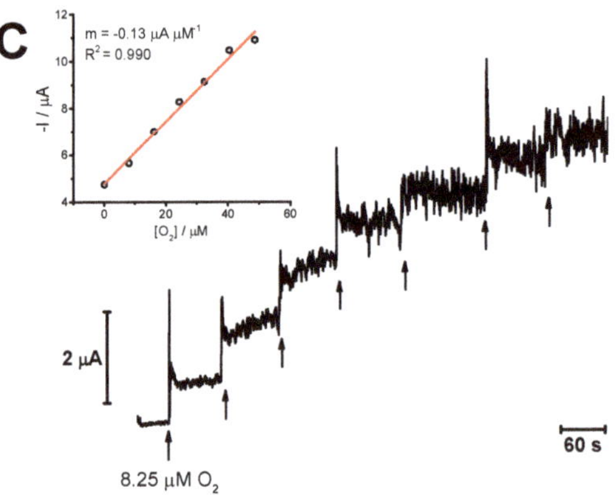

Figure 6. Electrochemical behavior of O_2 reduction at the depth probe surface. (**A**) Average sensitivity as a function of the applied potential, obtained from calibration of 4 sites in PBS. (**B**) Average baseline current as a function of the applied potential, obtained in N_2-purged PBS. (**C**) Representative calibration obtained at -0.6 V vs. Ag/AgCl and the calibration curve (inset) of a single recording site of the Integra probe. $n = 4$.

4. Discussion

Platinum surfaces are of paramount importance in applications such as those involving neural interfacing and design of electrochemical sensors and biosensors [24]. In the present study, we have investigated the electrochemical properties of a clinical depth recording electrode, in particular its suitability for monitoring pO_2 in brain tissue.

It is commonly accepted that $PbtO_2$ reflects changes in metabolism and cerebrovascular response and monitoring this parameter has become increasingly standardized in neurocritical care [33]. Currently available probes for monitoring $PbtO_2$ in situ in patients include the Clark-type amperometric Licox® probe and the Neurovent®-PTO optical probe, both of which are single site recording devices [21,22]. Based on our previous experience with multisite Pt microelectrode arrays, we propose that clinical multisite recording electrodes such as the depth probe used here may be suitable for monitoring $PbtO_2$ from multiple sites in the brain, offering a more integrated vision of brain oxygenation.

The morphological evaluation of the recording surface of the depth electrode revealed a smooth surface of Pt over what appears to be an Al-oxide substrate, most likely Al_2O_3. The Pt coverage was shown not to be complete, accounting for roughly 90% of the surface area. This was further corroborated by determination of the electroactive surface area, which indicated a roughness factor of 0.9.

The electrochemical characterization of the Pt surface performed by cyclic voltammetry in acidic media revealed redox peaks at -0.08 and -0.2 V, which could be attributed to strong and weak proton adsorption on Pt surfaces with (100) and (110) basal planes of a polycrystalline structure, respectively [34,35]. Furthermore, the presence of the three distinct peaks for H^+ desorption indicates a high-quality Pt surface [35,36]. An oxidation wave is observed for $E > 0.5$ V due to the formation of Pt oxide species Pt-O and Pt-OH, and there is a strong reduction peak at about 0.52 V corresponding to oxide reduction. Furthermore, the potential window of 1.5V did not widen in buffered aqueous solution corroborating the smooth structure of the Pt surface as nanostructure surfaces tend to show an increase in the potential width in neutral vs. acid media [35].

The electrochemical impedance spectroscopy revealed that the value of Z' was 77.4 ± 20.6 Ω at 1 kHz, a value that is slightly lower than that reported for a Pt/Ir depth electrode with similar recording surface area [37], and much lower than that reported in our previous study (0.2 MΩ) for a smooth thin-film Pt microelectrode array [26]. As impedance is inversely proportional to recording site size but should be low as to decrease noise in recording [25,38]. In microelectrodes, where high impedance due to size can become a limiting factor, surface area is typically increased through roughening or functionalization [39,40].

The evaluation of the analytical performance toward the oxygen reduction reaction revealed that the optimal working potential for monitoring O_2 is −0.6V vs. Ag/AgCl, in line with our previous observations. However, we found that the sensitivity was slightly lower than our previously reported value for thin-film Pt microelectrode arrays (3.2 ± 0.5 A M^{-1} cm^{-2}), while the LOD is in a similar range as previously reported value of 0.3 ± 0.2 µM [26]. This supports the suitability of the present probe to monitor changes in PbtO$_2$ in vivo in brain tissue, where basal values have been found to be approx. 30 µM [41]. We have previously shown that Pt electrode surfaces are optimal for the electrochemical reduction of O_2, allowing direct and real-time monitoring of PbtO$_2$ in brain tissue in rodent models [26]. Further validation of PbtO$_2$ monitoring in using this depth recording electrode will be carried out in a swine model.

Considering that brain cerebrovascular function and neurometabolism are coupled to neuronal activity, there is great interest in simultaneous monitoring of electrical activity and PbtO$_2$, as changes in PbtO$_2$ result from both cerebral blood flow and oxidative metabolism [42–44]. Currently available invasive technology requires that two separate probes be implanted in distinct brain regions [45], which greatly hinders the establishment of correlations between changes in tissue oxygenation and neural activity. In line with previous observations [46–48], we demonstrated that high frequency amperometry can concurrently report PbtO$_2$ and local field potential related currents in vivo in a rodent model of seizures [49]. Combined with other studies using enzyme-based biosensors and the same methodological approach, strong evidence supports that the high frequency component (>1 Hz) of an amperometric signal is a surrogate signal for local field potential [47,48,50]. This is an attractive approach toward simultaneous monitoring of electrical and chemical activity in vivo in brain tissue using a seamless methodology (amperometry) and a single probe that may have multiple recording sites [51].

5. Conclusions

We present an electrochemical evaluation of the analytical properties toward oxygen detection of a clinical grade depth recording electrode comprising 12 Pt recording contacts. Based on the reported catalytical properties of Pt toward the electroreduction reaction of O_2, we propose that these probes could be repurposed for multisite monitoring of PbtO$_2$ in vivo in the human brain. We found that the surface of the recording sites is composed of a thin film of smooth Pt and that the electrochemical behavior evaluated by cyclic voltammetry in acidic and neutral electrolytes is typical of a polycrystalline Pt surface. The smoothness of the Pt surface was further corroborated by determination of the electrochemical active surface, confirming a roughness factor of 0.9. At an optimal working potential of −0.6 V vs. Ag/AgCl, the sensor displayed suitable values of sensitivity and LOD, supporting its capability for monitoring PbtO$_2$ in vivo in the brain. The repurposing of these probes from electrophysiology to electrochemical detection of O_2 will allow seamless multisite monitoring of PbtO$_2$ in clinical setting, which holds promise in the context of multimodal monitoring in neurocritical care where PbtO$_2$ has become increasingly standardized following evidence of improved patient outcomes when PbtO$_2$ targeted therapeutic approaches are used.

Supplementary Materials: The following are available online at http://www.mdpi.com/2072-666X/11/7/632/s1. Figure S1: Photograph and detail of the Aurogen® probe used in the present study; Figure S2: SEM and EDS analysis of depth probe before and after electrochemical evaluation

Author Contributions: Conceptualization, A.L., G.A.G., and R.M.B.; methodology, A.L. and R.M.B.; formal analysis, A.L. and E.F.; investigation, A.L. and E.F.; resources J.E.Q. and G.A.G.; writing—original draft preparation, A.L.; writing—review and editing, J.E.Q., G.A.G., and R.M.B.; supervision, A.L. and R.M.B.; funding acquisition, R.M.B. and G.A.G. All authors have read and agreed to the published version of the manuscript.

Funding: This work was financed by the European Regional Development Fund (ERDF) through the COMPETE 2020 – Operational Programme for Competitiveness and Internationalization and Portuguese national funds via FCT – Fundação para a Ciência e Tecnologia, under projects POCI-01-0145-FEDER-028261 and UIDB/04539/2020.

Micromachines **2020**, *11*, 632

Acknowledgments: Additional support was provided through the University of Kentucky Brain Restoration Center and the Integra electrodes were provided in-kind by Integra LifeSciences.

Conflicts of Interest: The authors declare no conflict of interest.

References

1. Tasneem, N.; Samaniego, E.A.; Pieper, C.; Leira, E.C.; Adams, H.P.; Hasan, D.; Ortega-Gutierrez, S. Brain Multimodality Monitoring: A New Tool in Neurocritical Care of Comatose Patients. *Crit. Care Res. Pract.* **2017**, *2017*, 1–8. [CrossRef] [PubMed]
2. Scheeren, T.W.L.; Kuizenga, M.H.; Maurer, H.; Struys, M.M.R.F.; Heringlake, M. Electroencephalography and Brain Oxygenation Monitoring in the Perioperative Period. *Anesth. Analg.* **2019**, *128*, 265–277. [CrossRef] [PubMed]
3. Lang, E.W.; Jaeger, M. Systematic and comprehensive literature review of publications on direct cerebral oxygenation monitoring. *Open Crit. Care Med. J.* **2013**, *6*, 1–24. [CrossRef]
4. Bohman, L.E.; Pisapia, J.M.; Sanborn, M.R.; Frangos, S.; Lin, E.; Kumar, M.; Park, S.; Kofke, W.A.; Stiefel, M.F.; Leroux, P.D.; et al. Response of brain oxygen to therapy correlates with long-term outcome after subarachnoid hemorrhage. *Neurocrit. Care* **2013**, *19*, 320–328. [CrossRef]
5. Stiefel, M.F.; Spiotta, A.; Gracias, V.H.; Garuffe, A.M.; Guillamondegui, O.; Maloney-Wilensky, E.; Bloom, S.; Grady, M.S.; LeRoux, P.D. Reduced mortality rate in patients with severe traumatic brain injury treated with brain tissue oxygen monitoring. *J. Neurosurg.* **2005**, *103*, 805–811. [CrossRef]
6. Maas, A.I.; Fleckenstein, W.; de Jong, D.A.; van Santbrink, H. Monitoring cerebral oxygenation Experimental studies and preliminary clinical results of continuous monitoring of cerebrospinal fluid and brain tissue oxygen tension. *Acta Neurochir. Suppl. (Wien)* **1993**, *59*, 50–57. [CrossRef]
7. Rosner, M.J.; Zauner, A.; Bullock, M.R.R.; Levy, M.L. Clinical experience with 118 brain tissue oxygen partial pressure catheter probes: Comments. *Neurosurgery* **1998**, *43*, 1094–1095. [CrossRef]
8. Maas, A.I.R.; Fleckenstein, W.; de Jong, D.A.; Wolf, M. Effect of Increased ICP and Decreased Cerebral Perfusion Pressure on Brain Tissue and Cerebrospinal Fluid Oxygen Tension. In *Proceedings of the Intracranial Pressure VIII*; Avezaat, C.J.J., van Eijndhoven, J.H.M., Maas, A.I.R., Tans, J.T.J., Eds.; Springer: Berlin/Heidelberg, Germany, 1993; pp. 233–237.
9. Kiening, K.L.; Unterberg, A.W.; Bardt, T.F.; Schneider, G.H.; Lanksch, W.R. Monitoring of cerebral oxygenation in patients with severe head injuries: Brain tissue PO2 versus jugular vein oxygen saturation. *J. Neurosurg.* **1996**, *85*, 751–757. [CrossRef]
10. Bratton, S.L.; Chestnut, R.M.; Ghajar, J.; McConnell Hammond, F.F.; Harris, O.A.; Hartl, R.; Manley, G.T.; Nemecek, A.; Newell, D.W.; Rosenthal, G.; et al. Brain Oxygen Monitoring and Thresholds. *J. Neurotrauma* **2007**, *24*, S65–S70. [CrossRef]
11. Bhatia, A.; Gupta, A.K. Neuromonitoring in the intensive care unit. II. Cerebral oxygenation monitoring and microdialysis. *Intensive Care Med.* **2007**, *33*, 1322–1328. [CrossRef]
12. Salehi, A.; Zhang, J.H.; Obenaus, A. Response of the cerebral vasculature following traumatic brain injury. *J. Cereb. Blood Flow Metab.* **2017**, *37*, 2320–2339. [CrossRef]
13. Zhou, T.; Kalanuria, A. Cerebral Microdialysis in Neurocritical Care. *Curr. Neurol. Neurosci. Rep.* **2018**, *18*. [CrossRef] [PubMed]
14. Oddo, M.; Levine, J.M.; MacKenzie, L.; Frangos, S.; Feihl, F.; Kasner, S.E.; Katsnelson, M.; Pukenas, B.; MacMurtrie, E.; Maloney-Wilensky, E.; et al. Brain hypoxia is associated with short-term outcome after severe traumatic brain injury independently of intracranial hypertension and low cerebral perfusion pressure. *Neurosurgery* **2011**, *69*, 1037–1045. [CrossRef]
15. Dreier, J.P.; Fabricius, M.; Ayata, C.; Sakowitz, O.W.; William Shuttleworth, C.; Dohmen, C.; Graf, R.; Vajkoczy, P.; Helbok, R.; Suzuki, M.; et al. Recording, analysis, and interpretation of spreading depolarizations in neurointensive care: Review and recommendations of the COSBID research group. *J. Cereb. Blood Flow Metab.* **2017**, *37*, 1595–1625. [CrossRef]
16. Hartings, J.A. Spreading depolarization monitoring in neurocritical care of acute brain injury. *Curr. Opin. Crit. Care* **2017**, *23*, 94–102. [CrossRef]
17. Metzen, E.; Ratcliffe, P.J. HIF hydroxylation and cellular oxygen sensing. *Biol. Chem.* **2004**, *385*, 223–230. [CrossRef]

18. De Georgia, M.A. Brain Tissue Oxygen Monitoring in Neurocritical Care. *J. Intensive Care Med.* **2015**, *30*, 473–483. [CrossRef] [PubMed]
19. Murkin, J.M.; Arango, M. Near-infrared spectroscopy as an index of brain and tissue oxygenation. *Br. J. Anaesth.* **2009**, *103*, 3–13. [CrossRef]
20. Huschak, G.; Hoell, T.; Hohaus, C.; Kern, C.; Minkus, Y.; Meisel, H.-J. Clinical Evaluation of a New Multiparameter Neuromonitoring Device: Measurement of Brain Tissue Oxygen, Brain Temperature, and Intracranial Pressure. *J. Neurosurg. Anesthesiol.* **2009**, *21*, 155–160. [CrossRef] [PubMed]
21. Stewart, C.; Haitsma, I.; Zador, Z.; Hemphill, J.C.; Morabito, D.; Manley, G.; Rosenthal, G. The new licox combined brain tissue oxygen and brain temperature monitor: Assessment of in vitro accuracy and clinical experience in severe traumatic brain injury. *Neurosurgery* **2008**, *63*, 1159–1164. [CrossRef] [PubMed]
22. Dmitriev, R.I.; Papkovsky, D.B. Optical probes and techniques for O 2 measurement in live cells and tissue. *Cell. Mol. Life Sci.* **2012**, *69*, 2025–2039. [CrossRef] [PubMed]
23. Kirkman, M.A.; Smith, M. Brain Oxygenation Monitoring. *Anesthesiol. Clin.* **2016**, *34*, 537–556. [CrossRef] [PubMed]
24. Cowley, A.; Woodward, B. A healthy future: Platinum in medical applications platinum group metals enhance the quality of life of the global population. *Platin. Met. Rev.* **2011**, *55*, 98–107. [CrossRef]
25. Cogan, S.F. Neural Stimulation and Recording Electrodes. *Annu. Rev. Biomed. Eng.* **2008**, *10*, 275–309. [CrossRef]
26. Ledo, A.; Lourenço, C.F.; Laranjinha, J.; Brett, C.M.A.; Gerhardt, G.A.; Barbosa, R.M. Ceramic-Based Multisite Platinum Microelectrode Arrays: Morphological Characteristics and Electrochemical Performance for Extracellular Oxygen Measurements in Brain Tissue. *Anal. Chem.* **2017**, *89*, 1674–1683. [CrossRef]
27. Wu, J.; Yang, H. Platinum-based oxygen reduction electrocatalysts. *Acc. Chem. Res.* **2013**, *46*, 1848–1857. [CrossRef]
28. Sander, R. Compilation of Henry's law constants (version 4.0) for water as solvent. *Atmos. Chem. Phys.* **2015**, *15*, 4399–4981. [CrossRef]
29. Mocak, J.; Bond, A.M.; Mitchell, S.; Scollary, G. A statistical overview of standard (IUPAC and ACS) and new procedures for determining the limits of detection and quantification: Application to voltammetric and stripping techniques (Technical Report). *Pure Appl. Chem.* **1997**, *69*, 297–328. [CrossRef]
30. Bard, A.J.; Faulkner, L.R. Electrochemical methods: Fundamentals and applications. In *Electrochemical Methods: Fundamentals and Applications*; John Wiley & Sons, Inc: New York, NY, USA, 2000; ISBN 0-471-04372-9.
31. Licht, S.; Cammarata, V.; Wrighton, M.S. Direct measurements of the physical diffusion of redox active species: Microelectrochemical experiments and their simulation. *J. Phys. Chem.* **1990**, *94*, 6133–6140. [CrossRef]
32. Randles, J.E.B. Kinetics of rapid electrode reactions. *Discuss. Faraday Soc.* **1947**, *1*, 11. [CrossRef]
33. Vedantam, A.; Gopinath, S.P.; Robertson, C.S. Acute Management of Traumatic Brain Injury. In *Rehabilitation after Traumatic Brain Injury*; Elsevier: Amsterdam, The Netherlands, 2019; pp. 1–11.
34. Chen, D.; Tao, Q.; Liao, L.W.; Liu, S.X.; Chen, Y.X.; Ye, S. Determining the Active Surface Area for Various Platinum Electrodes. *Electrocatalysis* **2011**, *2*, 207–219. [CrossRef]
35. Daubinger, P.; Kieninger, J.; Unmüssig, T.; Urban, G.A. Electrochemical characteristics of nanostructured platinum electrodes-A cyclic voltammetry study. *Phys. Chem. Chem. Phys.* **2014**, *16*, 8392–8399. [CrossRef] [PubMed]
36. Inzelt, G.; Berkes, B.B.; Kriston, Á. Electrochemical nanogravimetric studies of adsorption, deposition, and dissolution processes occurring at platinum electrodes in acid media. *Pure Appl. Chem.* **2010**, *83*, 269–279. [CrossRef]
37. Klimes, P.; Duque, J.J.; Brinkmann, B.; Van Gompel, J.; Stead, M.; St. Louis, E.K.; Halamek, J.; Jurak, P.; Worrell, G. The functional organization of human epileptic hippocampus. *J. Neurophysiol.* **2016**, *115*, 3140–3145. [CrossRef]
38. Szostak, K.M.; Grand, L.; Constandinou, T.G. Neural Interfaces for Intracortical Recording: Requirements, Fabrication Methods, and Characteristics. *Front. Neurosci.* **2017**, *11*. [CrossRef]
39. Shokoueinejad, M.; Park, D.-W.; Jung, Y.; Brodnick, S.; Novello, J.; Dingle, A.; Swanson, K.; Baek, D.-H.; Suminski, A.; Lake, W.; et al. Progress in the Field of Micro-Electrocorticography. *Micromachines* **2019**, *10*, 62. [CrossRef] [PubMed]
40. Boehler, C.; Stieglitz, T.; Asplund, M. Nanostructured platinum grass enables superior impedance reduction for neural microelectrodes. *Biomaterials* **2015**, *67*, 346–353. [CrossRef]

41. Pennings, F.A.; Schuurman, P.R.; van den Munckhof, P.; Bouma, G.J. Brain Tissue Oxygen Pressure Monitoring in Awake Patients during Functional Neurosurgery: The Assessment of Normal Values. *J. Neurotrauma* **2008**, *25*, 1173–1177. [CrossRef]
42. Johnston, A.J.; Steiner, L.A.; Gupta, A.K.; Menon, D.K. Cerebral oxygen vasoreactivity and cerebral tissue oxygen reactivity. *Br. J. Anaesth.* **2003**, *90*, 774–786. [CrossRef]
43. Watts, M.E.; Pocock, R.; Claudianos, C. Brain Energy and Oxygen Metabolism: Emerging Role in Normal Function and Disease. *Front. Mol. Neurosci.* **2018**, *11*, 216. [CrossRef]
44. Schurr, A.; Miller, J.J.; Payne, R.S.; Rigor, B.M. An increase in lactate output by brain tissue serves to meet the energy needs of glutamate-activated neurons. *J. Neurosci.* **1999**, *19*, 34–39. [CrossRef] [PubMed]
45. Al-Mufti, F.; Lander, M.; Smith, B.; Morris, N.A.; Nuoman, R.; Gupta, R.; Lissauer, M.E.; Gupta, G.; Lee, K. Multimodality Monitoring in Neurocritical Care: Decision-Making Utilizing Direct And Indirect Surrogate Markers. *J. Intensive Care Med.* **2019**, *34*, 449–463. [CrossRef] [PubMed]
46. Zhang, H.; Lin, S.-C.; Nicolelis, M.A.L. Acquiring local field potential information from amperometric neurochemical recordings. *J. Neurosci. Methods* **2009**, *179*, 191–200. [CrossRef]
47. Santos, R.M.; Laranjinha, J.; Barbosa, R.M.; Sirota, A. Simultaneous measurement of cholinergic tone and neuronal network dynamics in vivo in the rat brain using a novel choline oxidase based electrochemical biosensor. *Biosens. Bioelectron.* **2015**, *69*, 83–94. [CrossRef] [PubMed]
48. Lourenço, C.F.; Ledo, A.; Gerhardt, G.A.; Laranjinha, J.; Barbosa, R.M. Neurometabolic and electrophysiological changes during cortical spreading depolarization: Multimodal approach based on a lactate-glucose dual microbiosensor arrays. *Sci. Rep.* **2017**, *7*. [CrossRef]
49. Ledo, A.; Lourenço, C.F.; Laranjinha, J.; Gerhardt, G.A.; Barbosa, R.M. Combined in vivo Amperometric Oximetry and Electrophysiology in a Single Sensor—A Tool for Epilepsy Research. *Anal. Chem.* **2017**, *89*, acs.analchem.7b03452 [CrossRef]
50. Viggiano, A.; Marinesco, S.; Pain, F.; Meiller, A.; Gurden, H. Reconstruction of field excitatory post-synaptic potentials in the dentate gyrus from amperometric biosensor signals. *J. Neurosci. Methods* **2012**, *206*, 1–6. [CrossRef]
51. Ledo, A.; Lourenço, C.F.; Laranjinha, J.; Gerhardt, G.A.; Barbosa, R.M. Concurrent measurements of neurochemical and electrophysiological activity with microelectrode arrays: New perspectives for constant potential amperometry. *Curr. Opin. Electrochem.* **2018**, *12*, 129–140. [CrossRef]

micromachines

Article

Mechanically Robust, Softening Shape Memory Polymer Probes for Intracortical Recording

Allison M. Stiller [1,*], Joshua O. Usoro [1], Jennifer Lawson [1], Betsiti Araya [1], María Alejandra González-González [2], Vindhya R. Danda [3], Walter E. Voit [1,3,4], Bryan J. Black [1] and Joseph J. Pancrazio [1]

[1] Department of Bioengineering, The University of Texas at Dallas, Richardson, TX 75080, USA; Joshua.usoro@utdallas.edu (J.O.U.); jxl125231@utdallas.edu (J.L.); betsiti.araya@utdallas.edu (B.A.); walter.voit@utdallas.edu (W.E.V.); bjb140530@utdallas.edu (B.J.B.); joseph.pancrazio@utdallas.edu (J.J.P.)

[2] Department of Biomedical Engineering, University of Houston, Houston, TX 77004, USA; magonzalezg@uh.edu

[3] Qualia, Inc., Dallas, TX 75252, USA; vindhya@qualiamedical.com

[4] Department of Materials Science and Engineering, The University of Texas at Dallas, Richardson, TX 75080, USA

* Correspondence: Allison.stiller@utdallas.edu

Received: 21 May 2020; Accepted: 23 June 2020; Published: 25 June 2020

Abstract: While intracortical microelectrode arrays (MEAs) may be useful in a variety of basic and clinical scenarios, their implementation is hindered by a variety of factors, many of which are related to the stiff material composition of the device. MEAs are often fabricated from high modulus materials such as silicon, leaving devices vulnerable to brittle fracture and thus complicating device fabrication and handling. For this reason, polymer-based devices are being heavily investigated; however, their implementation is often difficult due to mechanical instability that requires insertion aids during implantation. In this study, we design and fabricate intracortical MEAs from a shape memory polymer (SMP) substrate that remains stiff at room temperature but softens to 20 MPa after implantation, therefore allowing the device to be implanted without aids. We demonstrate chronic recordings and electrochemical measurements for 16 weeks in rat cortex and show that the devices are robust to physical deformation, therefore making them advantageous for surgical implementation.

Keywords: intracortical microelectrode arrays; shape memory polymer; softening; robust

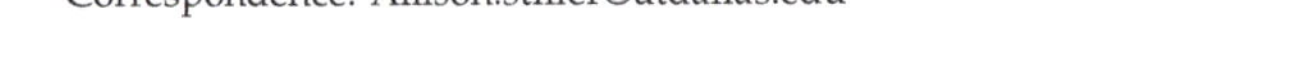

1. Introduction

Intracortical microelectrode arrays (MEAs) are devices that can be implanted in brain tissue to stimulate or record electrical activity from surrounding neural populations [1], making them important tools for investigating the function of the nervous system [2]. Additionally, intracortical recording MEAs have been used as critical components in brain–machine interfaces (BMIs) [3–5], which may be used to restore or replace loss of motor function in patients suffering from paralysis, limb loss, or neurodegenerative disorders. Despite the promise of MEA technology, the clinical adoption of these devices has been limited for many reasons, several of which are associated with stiff material composition. Commercially available devices are often fabricated from high modulus materials, such as silicon, to leverage reproducible photolithography techniques. However, this material choice also results in structures that are brittle due to the small device dimensions necessary to mitigate a severe chronic neuroinflammatory response, which may contribute to behavioral deficits [6]. This fragility renders devices vulnerable to fracture during fabrication [7] and handling [8,9]. Even after implantation, the device may be susceptible to breakage or cracking [10] due to the tethering forces caused by constant micromotion of the brain [11]. This effect may be exacerbated by the high degree of mechanical mismatch between the implanted device and the surrounding tissue, which creates a constant source of mechanical strain at the brain–device interface [12–14]. Because of this, many groups are investigating the potential use of robust, yet softer, polymer materials for intracortical device fabrication.

Unfortunately, there are many inherent issues associated with fabricating functional devices from soft materials. From a mechanical standpoint, polymer-based devices may lack the structural rigidity required to successfully implant in brain tissue if they are fabricated with small cross-sectional dimensions. To alleviate this issue, there have been demonstrations of the use of insertion guides to aid in implantation [15], but implementation of these aids may complicate surgical use. Others have addressed structural instability with stiffening coatings that dissolve after implantation [16–18], but these may increase the footprint of the device during implantation, thus increasing the risk of disrupting vasculature. Additionally, while many polymers such as SU-8 and Parylene C exhibit an elastic modulus of 1–5 GPa, approximately 2–3 orders of magnitude lower than that of materials commonly used for intracortical device fabrication, this is still 5–6 orders of magnitude higher than the estimated modulus of brain tissue (~0.5–10 kPa [19,20]). With these drawbacks in mind, there is motivation to design a robust polymer-based device that can be handled and implanted using existing surgical techniques, but that exhibits material properties that more effectively bridge mechanical mismatch with the brain.

Our group has previously investigated the use of shape memory polymers (SMPs) for fabrication of intracortical devices [21]. SMPs are a class of material that are unique in their ability to undergo dynamic changes in material properties in response to stimuli from the environment [22–24]. Specifically, we used a thiol-ene/acrylate formulation softened from 2 GPa in dry, room temperature conditions to ~300 MPa in wet, body-temperature conditions. This range ensured that devices could maintain structural stability during implantation, but soften by an order of magnitude in a few minutes after making contact with the brain tissue. This thiol-ene/acrylate SMP softens primarily due to an increase in temperature and minimally due to water absorption (<3% by weight). In this study, we used photolithography to create reproducible SMP based devices with the ability to soften to 20 MPa after implantation, an order of magnitude softer than our previously reported functional softening biolectronics. We demonstrate that these devices are capable of intracortical recording, show consistent electrical performance after substantial mechanical perturbation, and induce a tissue response consistent with similarly sized silicon shanks. Furthermore, we provide pilot data indicative of a reduced behavioral deficit associated with the SMP based devices.

2. Materials and Methods

2.1. Polymer Preparation and Device Fabrication

The SMP substrate was prepared as previously described [25] and material properties were determined using dynamic mechanical analysis (DMA), a methodology that applies cyclic, tensile loading to samples while changing the temperature in dry or aqueous conditions. DMA results indicated that the SMP exhibited an elastic modulus of 2 GPa in dry, room-temperature conditions and 20 MPa in soaked, body-temperature conditions. The pre-polymer solution was prepared using 1,3,5-triallyl-1,3,5-triazine-2,4,6 (1H,3H,5H)-trione (TATATO), trimethylolpropane tris(3-mercaptopropionate) (TMTMP), and (2-(3-mercaptopropionyloxy) ethyl) isocyanurate (TMICN) at molar ratios of 0.5:0.45:0.05, respectively, with 0.1 weight percent 2,2-dimethoxy-2-phenylacetophenone (DMPA), a photoinitiator. The solution was spun onto 4 inch silicon wafers (test grade, laser scribed, P-type, boron-doped wafers) at 600 rpm for 30 s and subsequently cured under UV light for 3 min at 254 nm and 1 h at 365 nm. The polymerized samples were then transferred to an oven for an overnight vacuum bake at 120 °C. This spin coating process yielded a 29 μm thick SMP layer on the silicon wafers. All processing, with the exception of SMP spin-coating and device release, was done in a Class 10,000 cleanroom at the University of Texas at Dallas (UTD).

All photomasks needed for device fabrication were designed using AutoCAD and fabricated in the above cleanroom using a Heidelberg DWL66 laser lithography system (Heidelberg, Germany). The SMP layer was cleaned with deionized water followed by a nitrogen dry and dehydration bake at 115 °C in a vacuum oven for 15 min to promote adhesion. A 1 μm layer of Parylene C was then deposited on the bottom SMP layer to act as an insulator, using a SCS PDS 2010 coater (Specialty Coating Systems, Indianapolis, IN, USA). Gold interconnects (5 μm wide) were patterned by liftoff lithography using nLOF photoresist and low stress silicon nitride (900 mT, 150 °C, 100 W with SiH_4, NH_3, H_3, and N_2 gases at 280, 4600, and 200 sccm, respectively). A 400 nm gold thin film was electro-beam evaporated on the wafer. A secondary 1 μm layer of Parylene C was deposited, and the wafers were patterned using standard lithography techniques and etched using oxygen plasma to form a coaxial layer around the gold traces. The wafers then went through a dehydration bake at 115 °C for 15 min and immediately received a 7 μm thick top coat of the SMP using spin parameters of 1200 rpm for 2 min. They were then returned to the cleanroom for additional processing. Electrode sites, bond pads, and the device outline were patterned using standard photolithography methods. A hard mask of ~200 nm low stress silicon nitride was used to perform the plasma etching of SMP to open the electrode sites and bond pads and create the device outline. After etching, the hard mask was stripped off using hydrofluoric acid, and electrode sites (18 × 10 μm) were patterned with liftoff lithography using AZ400K solvent at room temperature. Next, the electrode sites received sputtered iridium oxide film (SIROF) to decrease electrode site impedance. First, a 20–50 μm layer of titanium was deposited (200 V, 4 Torr, with Ar flow at 50 sccm) to act as an adhesion layer for the SIROF. Next, a 160–180 μm layer of iridium oxide was sputtered onto the electrode sites (100 V, 30 Torr, with Ar, O_2, and 2% H_2/98% Ar at 15, 20, and 10 sccm, respectively). After the fabrication was complete, wafers were soaked in DI water in a 37 °C oven until individual devices released from the wafers. Released devices were assembled with packaging with Omnetics connectors (A79040-001, Omnetics Connectors Corporation, Minneapolis, MN, USA) by Qualia Labs, Inc. (Figure 1a). A thicker variation of the SMP devices were also fabricated for the purposes of mechanical testing. In this case, all fabrication steps were identical with the exception of a 29 μm thick top SMP layer in place of the 7 μm layer (Figure 1b). Size-matched non-functional silicon devices were also fabricated in UTD cleanroom facilities using standard photolithography techniques (Figure 1c).

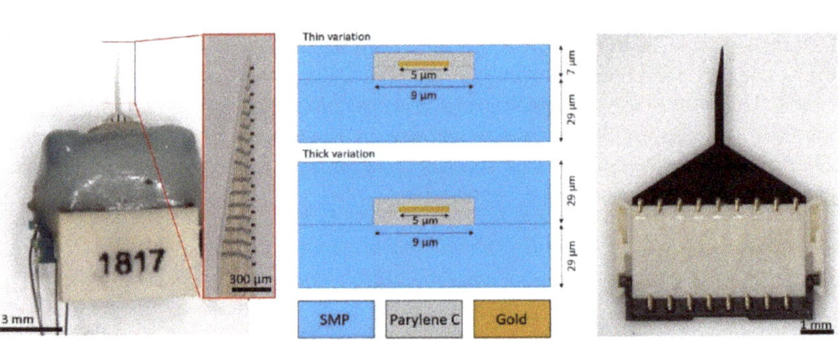

(a) (b) (c)

Figure 1. SMP and silicon device design. (**a**) Optical image of a packaged SMP device with the red inset depicting 16 electrode sites on an SMP device (the electrodes appear to be black due to the SIROF coating). The Omnetics connector is sealed off with epoxy (grey encapsulation material). (**b**) Representative cross-sectional view of a gold trace and coaxial Parylene C insulation in both device variations. (**c**) A size-matched non-functional silicon device mounted in a zero insertion force (ZIF) connector.

2.2. Surgical Implantation

All animal handling, housing, and surgical procedures were approved by the University of Texas Institutional Animal Care and Use Committee. Male, Long Evans rats (n = 8, Charles River) weighing 600–800 g received bilateral implants of functional SMP devices and non-functional silicon shanks.

Surgical methods followed those previously reported [21]. Briefly, animals were anesthetized with an intraperitoneal injection of a ketamine cocktail (ketamine (65 mg/kg), xylazine (13.33 mg/kg), and acepromazine (1.5 mg/kg)) followed by an intramuscular injection of atropine sulfate (0.05 mg/kg) to counteract depressive cardiac effects of the ketamine and a subcutaneous injection of dexamethasone (2 mg/kg) to help mitigate post-operative inflammation. When unconscious, the animal's head was shaved and then mounted in a stereotaxic frame with a nose cone delivering 2% isoflurane in 100% oxygen for maintenance of the anesthetic plane. Ophthalmic ointment was applied to the eyes to retain moisture and prevent drying during the surgery. The scalp was cleaned alternating with 10% iodine solution and 70% ethanol wipes three times and then injected with 0.4 mL of 0.5% lidocaine to numb the area. An incision was made using a surgical blade down the midline of the scalp, and hemostatic forceps were used to hold the skin away from the skull. The surface of the skull was thoroughly cleaned and roughened to promote adhesion of the head cap.

First, a surgical drill was used to create three small holes in the skull for positioning of stainless steel screws. These screws provided mechanical support for the head cap but also served as ground and reference connections for external wiring on the functional SMP device. Next, 1–2 mm craniotomies were drilled over each motor cortex, centered approximately 2.5 mm rostral and 2.5 mm lateral from bregma (Figure 2a). The dura was resected over one craniotomy and a device was positioned over the center (Figure 2b). The device was implanted at a speed of 1 mm/s using a hydraulic micropositioner (Kopf Instruments, Tujunga, CA, USA) to a depth of 2 mm. The dura was then replaced with collagen-based grafts (Biodesign Dural Graft, Cook Medical, Bloomington, IN, USA). After the grafts were rehydrated by the surrounding cerebrospinal fluid, a layer of Gluture (World Precision Instruments, Sarasota, FL, USA), a biocompatible cyanoacrylate, was used to seal off the craniotomy. After implantation of the SMP device, ground and reference wires were wrapped around stainless steel screws while waiting for the cyanoacrylate to dry. Finally, a small amount of dental cement was applied around the device base in order to stabilize it during implantation of the second device. This process was then repeated for the second device in the contralateral cortex. Following both implantations, the head cap was built up to encapsulate the devices, leaving only the Omnetics connector visible. The skin behind the implantation sites was closed using surgical staples and left in place for 10 days following the operation. The animal then received 2–3 mL of sterile saline for hydration, a subcutaneous injection of sustained release buprenorphine SR LAB (ZooPharm, Windsor, CO, USA), and an intramuscular injection of cefazolin antibiotic (5 mg/kg). The animal received a follow-up shot of buprenorphine 72 h after surgery.

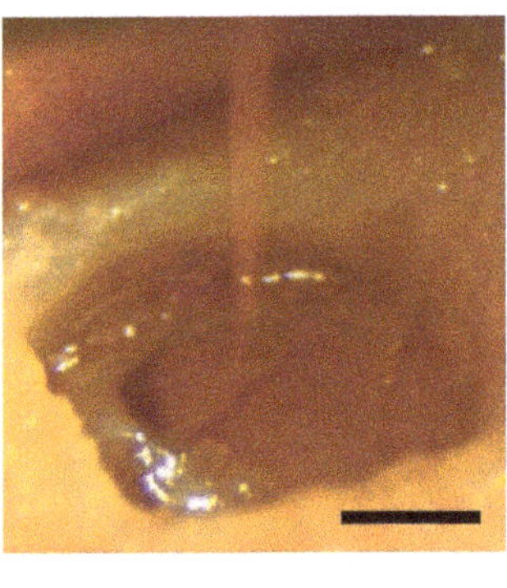

(a)

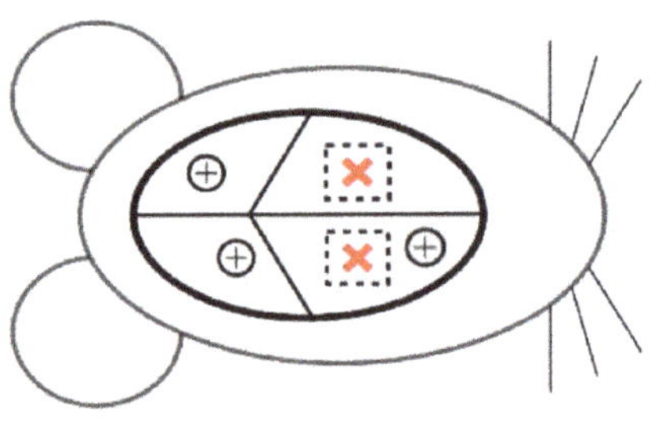

(b)

Figure 2. Device implantation. (**a**) Schematic depicting bilateral implant locations (red crosses) in rat brain. The circled plus signs indicate placement of stainless steel screws. (**b**) SMP shank held over a craniotomy with resected dura immediately before implantation. Scale bar = 1 mm.

2.3. Electrophysiological Recordings and Electrochemistry

We performed weekly electrophysiological recordings and electrochemistry on anesthetized animals (1–2% isoflurane) starting immediately after surgery, at Week 0, for 16 weeks. Spontaneous wideband data were recorded for 10 min at 40 kHz with functional SMP devices using a 16-channel Plexon headstage and Omniplex acquisition system (Plexon, Inc., Dallas, TX, USA). A four-pole Butterworth high pass filter with a cutoff frequency of 250 Hz was applied to the wideband data, and spikes were manually sorted using 2D principal component space after identifying potential spikes that crossed a threshold of −4 σ (based on the root mean square (RMS) of the filtered signal). Units were only considered for analysis if they contained at least 100 spike waveforms. A custom MATLAB code was used to extract mean peak-to-peak voltage (Vpp), RMS noise, and signal to noise ratio (SNR) calculated by dividing Vpp by two times the RMS noise of the associated channel.

Electrochemical impedance spectroscopy (EIS) was performed using a model 604E series electrochemical analyzer/workstation (CH Instruments, Inc., Austin, TX, USA). An 18-pin dual strip Nano-D female connector (NSD-18-WD-18.0-C-GS, Omnetics Connector Corporation, Minneapolis, MN, USA) was attached to the SMP Omnetics connector to interface with the system. The device wires wrapped around the stainless steel screws were shorted together to act as a reference electrode and the SIROF-coated electrodes served as the working electrodes. A 10 mV RMS sinusoidal signal ranging from 100 kHz to 1 Hz was sent to the device, and current was recorded 3 times per decade of frequency. CH Instruments software calculated impedance magnitude, and a custom MATLAB code extracted real impedance values at each point.

2.4. Behavioral Testing

In an effort to compare the potential behavioral changes associated with implantation of SMP and silicon shanks, we conducted pilot experiments using the cylinder test. The cylinder test takes advantage of rodents' natural affinity towards exploring novel environments to determine changes in limb-use asymmetry and overall exploratory behavior over time [26,27]. Rats were placed in a transparent plastic cylinder and were allowed to freely explore for 3 min while being filmed from underneath through a clear acrylic sheet (Figure 3a). The number of times the left paw, right paw, or both paws, were used to initially reach out and touch the walls of the cylinder was manually counted from the video (Figure 3b). Initial paw preference was determined by baseline measurements that were taken prior to surgical implantation. Cylinder testing was performed weekly beginning one week post-implantation. For analysis, rats were grouped based on initial paw preference and the type of device that was implanted contralateral to that paw. The percent use of each paw was then tracked over the 16-week study.

(a)

(b)

Figure 3. Cylinder test setup. (**a**) The rat is filmed from underneath as it rears up. (**b**) The rat rears up to explore its surroundings. The yellow circle indicates a paw touch against the cylinder.

2.5. Immunohistochemistry and Analysis

After 16 weeks, animals were sacrificed with an intraperitoneal injection of sodium pentobarbital (200 mg/kg). Tissue fixation was performed by transcardial perfusion with 4% paraformaldehyde (PFA). Whole brains were dissected, sectioned into hemispheres, and stored in PBS + 0.5% sodium azide at 4 °C. For cryoprotection, brains were incubated overnight in 15% sucrose (Sigma-Aldrich, St. Louis, MO, USA) solution in PBS and the following night in a 30% sucrose solution. The samples were then transferred to cryomolds filled with optimal cutting temperature compound (OCT, Sekura Finetek, Torrance, CA, USA). Flash freezing was then achieved by submerging cryomolds containing individual hemispheres into dry ice plus isopropyl alcohol. Once the OCT-embedded samples were frozen, they were placed in a −20 °C freezer until sectioned. Individual brain hemispheres were horizontally sectioned into 50 μm slices using a Leica CM3050 S cryostat and adsorbed onto charged slides (Fisher Scientific, Waltham, MA, USA) for staining.

Cryosections were permeabilized using 0.5% Triton X-100 (Sigma-Aldrich, USA) in 4% normal goat serum (NGS) (Abcam, Cambridge, UK) and 1× PBS and then blocked in a solution of 0.5% TritonX, 4% NGS, and 1× PBS. Sections were then incubated with mouse anti-neuronal nuclei (NeuN, 1:200, Abcam, ab104224), chicken anti-glial fibrillary acidic protein (GFAP, 1:200, Abcam, ab4674), and rabbit anti-CD-68 (1:200, Abcam, ab125212) diluted in a solution of 4% NGS and PBS for 24 h on a rocker at 4 °C. Following primary incubation, tissue was washed 3 times for 30 min using PBS + 0.5% Triton X-100 solution and then washed with 4% NGS + PBS solution for three hours. Sections were then incubated overnight on a rocker at 4 °C with secondary antibodies: goat anti-chicken Alexa Fluor 647 (1:200, Abcam, ab150171), goat anti-rabbit Alexa Fluor 488 (1:200, Abcam, ab150077), and goat anti-mouse Alexa Fluor 555 (1:200, Abcam, ab150118), along with 0.06% concentration DAPI (0.06%, Abcam, ab228549). Sections were then washed 3 times for 10 min in PBS solution. The PBS was replaced with drops of Image-iT (Invitrogen, Camarillo, CA, USA), and then sections were sealed with coverslips using clear nail polish.

Stained slides were imaged at 10× magnification on an inverted Nikon eclipse Ti (Nikon, Tokyo, Japan) scanning confocal microscope. With the implantation site centered in the field of view, 4-color z-stack confocal images were collected across the entire tissue slice thickness in 5 μm axial steps. Care was taken to adjust laser line and sensory gains such that no saturating pixels were detected. Each laser line was scanned individually to reduce signal bleed-over. Z-stack images were then converted to a single maximum intensity projection image for analysis. Normalized GFAP and CD68 fluorescence intensities and normalized NeuN+ cell counts were carried out using boutique ImageJ [28,29] macros, as previously described [30] with minor alterations. Briefly, any regions of the image that were dark due to absence of tissue (including tears or edges) were manually excluded and converted to "not a number" (NaN) values. Shank sites were then manually defined to enable analysis within concentric contours from the site. Average mean intensity values were then automatically calculated within concentric bands of 50 μm up to 400 μm from the shank site. Lastly, based on user-defined noise tolerances, a 3 pixel Gaussian blur was applied to the NeuN channel image and cell counts were automatically collected based on local maxima detection. GFAP intensity measurements and NeuN+ cell counts were then normalized to the furthest concentric band (350–400 μm).

2.6. Device Physical Robustness

We tested physical robustness of functional SMP devices compared to non-functional silicon devices using electrochemical endpoints after physical deformation. For these experiments, we tested an updated version of the device that featured a thickness of 58 μm but kept all other parameters constant. These devices were also tested in a subset of animals to demonstrate improved mechanical stability during implantation as compared to the thinner design. Before device bending, we performed EIS on functional SMP devices in 1× PBS, pH 7.4, at room temperature using methods identical to those described in Section 2.3. After testing, the devices were dipped in deionized water to remove any salts that accumulated on the electrode sites during testing and allowed to dry for 24 h.

Next, non-functional silicon devices were mounted on a stage such that the shank was parallel to the ground (Figure 4). The tip of the shank was then deflected at a speed of 1 mm/s using a custom attachment mounted to a NeuralGlider cortical neural implant inserter (Actuated Medical, Inc., Bellefonte, PA, USA). As soon as the shank broke, the system was manually stopped, and the displacement distance was recorded. This was repeated for n = 3 samples, and the average distance was recorded and used as an endpoint for deflection of the SMP devices. After drying completely, the SMP devices were mounted in an identical fashion to the silicon devices. The shank tips were displaced to the distance at which the silicon shanks broke at a speed of 1 mm/s. The SMP devices then underwent another round of EIS testing to allow for comparison of impedance magnitude values before and after deformation. Impedance magnitudes before and after physical deformation were then compared. Electrodes with an impedance magnitude greater than 1 MΩ at 1 kHz (approximately double the expected impedance) before deflection testing were excluded from analysis.

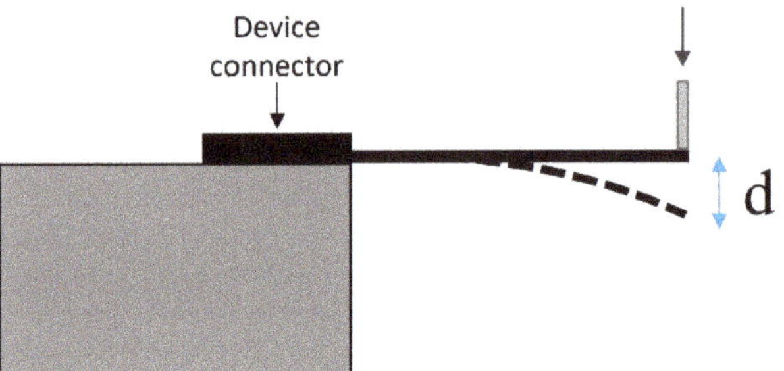

Figure 4. Device robustness testing setup. The device is mounted via the connector as a cantilevered beam on a stage with the entire length of the shank suspended parallel to the ground. The tip of the device is deflected downward at a speed of 1 mm/s (indicated by downward arrow) and the shank deforms accordingly (dashed line). In the case of the silicon shank, fracture occurs after a certain displacement (d). This displacement is then transiently applied in the same manner to the SMP devices.

2.7. Statistics

All statistics were performed in MATLAB R2018b (The MathWorks, Inc., Natick, MA, USA) and OriginPro 2020b (Origin Lab, Northampton, MA, USA). Significant increases/decreases ($p < 0.05$) in electrophysiological and electrochemical data over time were determined using analysis of variance (ANOVA) tests on residuals. ANOVA tests were performed to determine significant differences ($p < 0.05$) in immunohistological outcomes between brain slices containing silicon and SMP probes. A paired t-test was used to determine significant difference ($p < 0.05$) between EIS measurements taken before and after SMP probe deflection during robustness testing.

3. Results

3.1. Single Unit Recordings and In Vivo Electrochemistry

We performed weekly electrophysiological recordings from functional SMP devices in rat cortex and tracked single-unit amplitude, active electrode yield percentage (AEY), RMS noise, and signal-to-noise ratio (SNR) over 16 weeks (Figure 5). Results showed stable measurements for single-unit amplitude (Figure 5c), a decrease in RMS noise (Figure 5e), and an increase in signal-to-noise ratio over time (Figure 5f). However, the percentage of active electrodes, while relatively stable for the first 12–13 weeks, showed a decline in the last month of recording (Figure 5d). While one device failed to record single units during every recording session, 7 out of 8 devices exhibited active electrodes during each recording session.

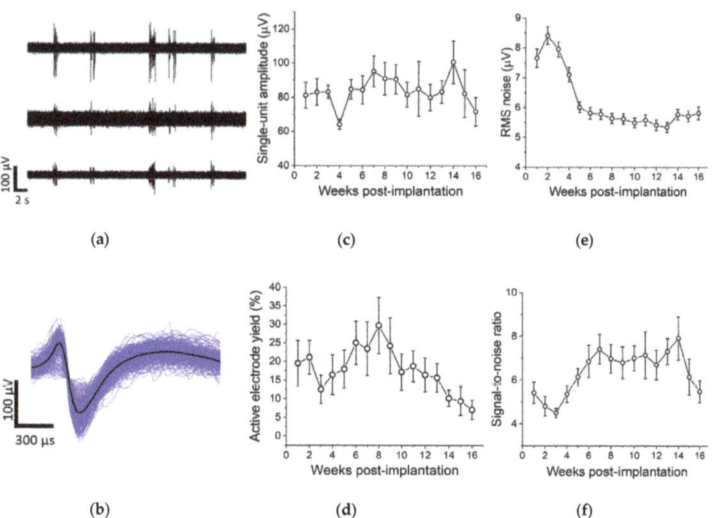

Figure 5. Chronic electrophysiological recordings. (**a**) Representative filtered data from three electrodes on one device. (**b**) Representative single unit activity on one electrode. We extracted measurements for single-unit amplitude (**c**), active electrode yield (**d**), RMS noise (**e**), and signal-to-noise ratio (**f**) from filtered wideband data. Linear regression analysis showed no change in single-unit amplitude over time. There was a decrease in active electrode yield ($R^2 = 0.03$, $p = 0.03$) and RMS noise ($R^2 = 0.09$, $p < 0.01$) and an increase in SNR ($R^2 = 0.02$, $p < 0.01$) over time. Data are represented as mean ± SEM.

We also performed weekly EIS measurements to assess electrode electrochemical stability. Results showed a slight decrease in impedance magnitude 1 kHz over 16 weeks with average values around 600 kΩ on Week 0 and 500 kΩ on Week 16 (Figure 6). Because of the ultra-softening nature of this SMP, we did not perform in vitro EIS on devices before implantation to premature material deformation. However, EIS measurements were taken on a small cohort of devices from the same fabrication batch to ensure impedance magnitudes were within an acceptable range (100–900 kΩ) on a majority of electrodes. In vitro measurements yielded impedance magnitudes (at 1 kHz) of ~100–250 kΩ, indicating that electrode impedance increased immediately after implantation.

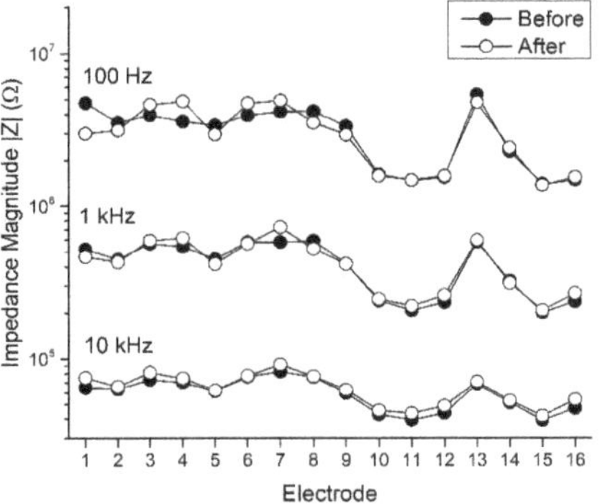

Figure 6. Impedance magnitude at 1 kHz over 16 weeks. Linear regression analysis shows a slight decrease in impedance over time ($R^2 = 0.03$, $p < 0.01$). Data are represented as mean ± SEM.

3.2. Device Physical Robustness

We subjected non-functional silicon shanks to deflection until fracture and then applied the same single displacement of approximately 1.92 mm to functional SMP devices. The mechanical deformation caused a minor, yet statistically significant change in the impedance magnitude at 1 kHz across the microelectrode sites of all tested devices. The mean impedance magnitude before deformation was 539.6 ± 39.0 kΩ, rising to 583.8 ± 46.9 kΩ following deformation, a 7.6% increase ($n = 31$ microelectrode sites, $p = 0.013$). Similar results were observed at 10 kHz, which showed 7.5 ± 1.2% ($p < 0.001$) increases after deformation. Increases in impedance at 100 Hz, 6.9 ± 3.9% were not statistically significant. Regardless, the electrochemical characteristics of the SMP microelectrode sites exhibited tolerance to mechanical deformation (Figure 7).

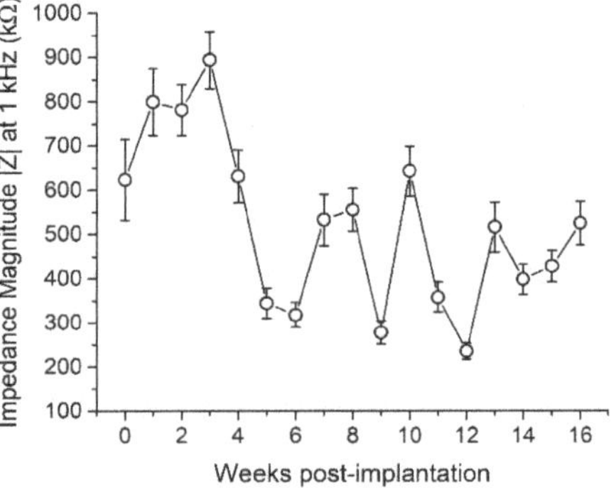

Figure 7. Representative EIS results from a single SMP device before and after deformation show minimal changes in impedance magnitude at 100 Hz, 1 kHz, and 10 kHz due to device deformation. The electrodes on this device exhibited an average decrease in impedance magnitude of $-0.05 \pm 4.2\%$ (mean ± SEM) at 100 Hz ($n = 16$, $p = 0.84$), and an average increase in impedance magnitude of 2.8 ± 2.4% at 1 kHz ($n = 16$, $p = 0.46$) and 7.2 ± 1.2% at 10 kHz ($n = 16$, $p < 0.001$). The electrode layout was such that electrode 1 was located at the tip of the shank and electrode 16 was furthest up the shank.

3.3. Immunohistochemistry

To assess the extent of neuroinflammation after 16 weeks, we performed immunohistochemistry to stain for markers associated with three cell types commonly implicated in the chronic neuroinflammatory response: GFAP for astrocytes, NeuN for neuronal nuclei, and CD68 for activated microglia/macrophages (Figure 8). We also measured the size of the hole left behind by the explanted device in each slice. Histological results were quantified for silicon and SMP devices at three depths along the device shank. Intensity or density values for each band were normalized to the band 350–400 µm from the perimeter of the device. ANOVA results indicated no significant differences in GFAP intensity, neuronal density, and CD68 intensity between silicon and SMP shanks at any comparison point (band or depth). There was also no significant difference in hole size between slices with silicon and SMP shanks at each depth; however, the hole size at the top layer (for both device types) was significantly larger than that found at the middle or deep slices. This trend was expected due to the tapered geometry of the shanks. Similarly, for both GFAP and CD68 intensity, we observed a trend of decreased astrogliosis and microglia/macrophage activation deeper along the shank, likely also in response to decreasing cross-section of the device. There appeared to be no loss of neuronal density in any band at any depth. Separate brain slices were also stained for immunoglobulin G (IgG), a marker associated with blood–brain-barrier leakage, but ANOVA results indicated no significant differences at any comparison point.

GFAP NeuN CD68 DAPI

Silicon SMP

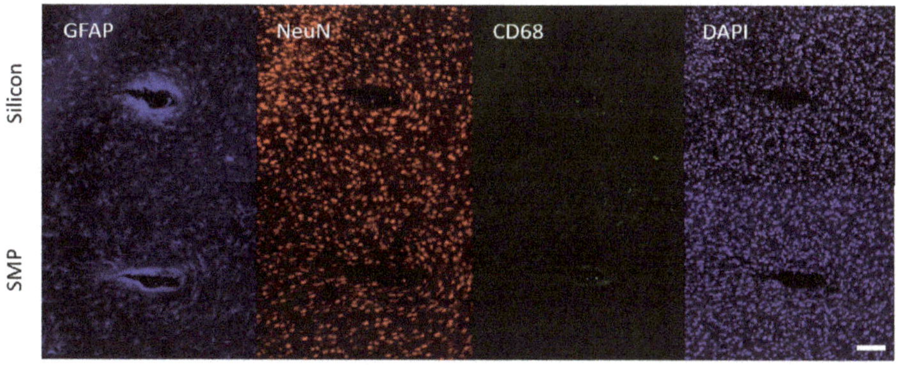

Figure 8. Representative immunofluorescent images for GFAP, NeuN, CD68, and DAPI stain in a middle brain slice containing a silicon device (top row) and an SMP device (bottom row). Scale bar = 100 µm.

3.4. Pilot Behavioral Deficit Data

All animals were assessed weekly for motor deficit through use of the cylinder test. Their initial preferred paw was determined by majority percentage of paw contacts from baseline data taken a week before device implantation. Figure 9 demonstrates percentage of paw contacts as a function of device type implanted contralateral to the preferred paw. Unfortunately, due to randomized placement of devices, there was only one animal with a silicon device implanted contralateral to its preferred paw. However, these preliminary data suggest that animals with the SMP device implanted contralateral to their preferred paw continued using this paw, whereas the opposite was true for the animal with a silicon shank implanted contralateral to the preferred paw.

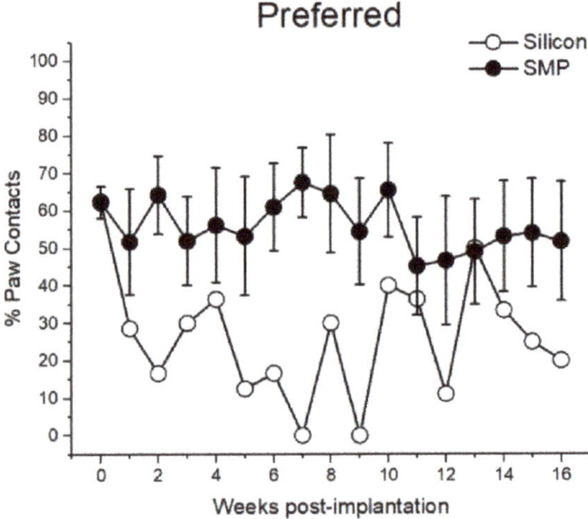

Figure 9. Percent of paw contacts as a function of device type implanted contralateral to the preferred paw (based on baseline data). SMP probe data are represented as mean ± SEM.

4. Discussion

Functional 16-channel SMP devices were implanted in rat motor cortexes for 16 weeks and their performance was tracked with weekly electrophysiological and electrochemical measurements. Within the first few weeks of implantation, active electrode yield decreased and impedance magnitude at 1 kHz increased; however, both of these trends began to reverse by week 4. Similar results were seen in our prior work [21] and may suggest onset and subsequent resolution of the acute neuroinflammatory response. While impedance magnitude remained stable for the remainder of the study, active electrode yield began dropping off in the last month of recordings. Though stable impedance values indicate that the integrity of the recording sites was likely maintained, it is possible that some degree of polymer breakdown could have contributed to decreased active electrode yield. Previous in vitro testing has validated that this variation of SMP is not hydrolytically stable and may experience degradation after chronic soaking [31]. Because of this, we are investigating a novel SMP formulation that is hydrolytically stable [31] for future chronic studies.

These updated devices were also used to demonstrate mechanical robustness in bend tests. SMP devices were able to withstand physical deformations that resulted in brittle fracture of silicon counterparts, a common issue that can make them difficult to handle during fabrication, packaging, and sterilization, and during surgical implantation. Polymers provide a clear advantage in that they may undergo large deformations but maintain integrity of electrical contacts. Furthermore, SMP-based devices may exhibit an even greater robustness to physical deformation when in the softened state due to the increase in flexibility, or decrease in stiffness, of the device. For example, the approximate stiffness of all discrete device components (gold electrode traces, Parylene C encapsulation, and SMP substrate) can be calculated using the modulus, area moment of inertia, and length of the individual components. SMP in the dry state exhibits a stiffness of approximately 1 N/m, but decreases to 0.01 N/m after softening. For comparison, an individual gold trace and coaxial Parylene C insulation exhibits a stiffness of approximately 10^{-7} (N/m) and 10^{-6} (N/m), respectively, meaning that neither the gold nor the Parylene C should inhibit robustness to physical deformation. The improved flexibility of this SMP formulation compared to that which was previously investigated (0.01 N/m versus 0.1 N/m, respectively) [21] is notable, considering the role that device stiffness may play in chronic functionality. The brain experiences continuous micromotion [11], thus warranting the need for a device that can endure constant mechanical disturbances. This need extends even further to brain macromotion, which may require even more flexibility of devices, especially in larger animal models or humans.

Additionally, given that SMP devices elicit a minimal neuroinflammatory response that is no worse than size-matched silicon devices, they may be advantageous alternatives devices for chronic use. Finally, preliminary behavioral outcomes indicated that softer devices might mitigate motor defects associated with implantation of stiff, silicon-based devices. More comprehensive analyses are required to examine the benefit of softening devices relative to behavioral measures.

5. Conclusions

In this work, we demonstrated fabrication and implementation of an SMP-based, 16-channel intracortical device that soften from ~2 GPa to ~20 MPa after implantation. We demonstrated an active electrode yield of 20–30% within the first 3 months of recordings with a final active electrode yield of ~10%. Weekly EIS measurements showed electrochemical stability over 16 weeks. Additionally, chronic immunohistological outcomes showed that SMP devices elicited a response no worse than bilaterally implanted silicon devices, making them good candidates for surgical implementation. Moreover, deformation testing indicated that SMP devices are mechanically robust as compared to silicon counterparts that are vulnerable to brittle fracture, further indicating that SMP offers material advantages for fabrication and handling.

Author Contributions: Conceptualization, A.M.S., B.J.B., W.E.V., and J.J.P.; methodology, A.S., J.O.U., J.L., M.A.G.-G., V.R.D., and B.J.B.; software, A.M.S., J.O.U., and B.J.B.; validation, A.M.S. and B.J.B.; formal analysis, A.M.S., J.O.U., and B.J.B.; investigation, A.M.S., J.O.U., B.A., J.L., and B.J.B.; resources, J.J.P.; data curation, A.M.S., J.O.U., B.A., J.L., and B.J.B.; writing—original draft preparation, A.M.S.; writing—review and editing, A.M.S., J.O.U., J.L., B.A., M.A.G.-G., V.R.D., B.J.B., J.J.P., and W.E.V.; visualization, A.M.S., J.O.U., and B.J.B.; supervision, J.J.P.; project administration, A.M.S. and J.J.P.; funding acquisition, J.J.P. and W.E.V. All authors have read and agreed to the published version of the manuscript.

Funding: This research was funded by the Office of the Assistant Secretary of Defense for Health Affairs through the Peer-Reviewed Medical Research Program, grant number W81XWH-15-1-0607, DARPA Phase II SBIR, grant number D17PC0011, and the Eugene McDermott Graduate Fellowship Program, grant number 201606 at The University of Texas at Dallas.

Conflicts of Interest: The authors declare no conflict of interest. The funders had no role in the design of the study; in the collection, analyses, or interpretation of data; in the writing of the manuscript, or in the decision to publish the results.

References

1. Ferguson, M.; Sharma, D.; Ross, D.; Zhao, F. A critical review of microelectrode arrays and strategies for improving neural interfaces. *Adv. Healthc. Mater.* **2019**, *8*. [CrossRef]
2. Guo, L. Principles of functional neural mapping using an intracortical ultra-density microelectrode array (ultra-density MEA). *J. Neural Eng.* **2020**. [CrossRef]
3. Hochberg, L.R.; Bacher, D.; Jarosiewicz, B.; Masse, N.Y.; Simeral, J.D.; Vogel, J.; Haddadin, S.; Liu, J.; Cash, S.S.; van der Smagt, P.; et al. Reach and grasp by people with tetraplegia using a neurally controlled robotic arm. *Nature* **2012**, *485*, 372–375. [CrossRef]
4. Tsu, A.P.; Burish, M.J.; GodLove, J.; Ganguly, K. Cortical neuroprosthetics from a clinical perspective. *Neurobiol. Dis.* **2015**, *83*, 154–160. [CrossRef]
5. Lebedev, M.A.; Nicolelis, M.A.L. Brain-machine interfaces: Past, present and future. *Trends Neurosci.* **2006**, *29*, 536–546. [CrossRef]
6. Goss-Varley, M.; Dona, K.R.; McMahon, J.A.; Shoffstall, A.J.; Ereifej, E.S.; Lindner, S.C.; Capadona, J.R. Microelectrode implantation in motor cortex causes fine motor deficit: Implications on potential considerations to Brain Computer Interfacing and Human Augmentation. *Sci. Rep.* **2017**, *7*. [CrossRef]
7. Du, J.; Roukes, M.L.; Masmanidis, S.C. Dual-side and three-dimensional microelectrode arrays fabricated from ultra-thin silicon substrates. *J. Micromech. Microeng.* **2009**, *19*. [CrossRef]
8. Schjetnan, A.G.P.; Luczak, A. Recording large-scale neuronal ensembles with silicon probes in the anesthetized rat. *J. Vis. Exp.* **2011**, *56*, 3282. [CrossRef]

9. Grand, L.B. Development, Testing and Application of Laminar Multielectrodes and Biocompatible Coatings for Intracortical Applications. Ph.D. Thesis, Pazmany Peter Catholic University, Budapest, Hungary, 2007.

10. Kozai, T.D.Y.; Catt, K.; Li, X.; Gugel, Z.V.; Olafsson, V.T.; Vazquez, A.L.; Cui, X.T. Mechanical failure modes of chronically implanted planar silicon-based neural probes for laminar recording. *Biomaterials* **2015**, *37*, 25–39. [CrossRef]

11. Gilletti, A.; Muthuswamy, J. Brain micromotion around implants in the rodent somatosensory cortex. *J. Neural Eng.* **2006**, *3*, 189–195. [CrossRef]

12. Lee, H.; Bellamkonda, R.V.; Sun, W.; Levenston, M.E. Biomechanical analysis of silicon microelectrode-induced strain in the brain. *J. Neural Eng.* **2005**, *2*, 81–89. [CrossRef] [PubMed]

13. Sridharan, A.; Nguyen, J.K.; Capadona, J.R.; Muthuswamy, J. Compliant intracortical implants reduce strains and strain rates in brain tissue in vivo. *J. Neural Eng.* **2015**, *12*. [CrossRef]

14. Subbaroyan, J.; Martin, D.C.; Kipke, D.R. A finite-element model of the mechanical effects of implantable microelectrodes in the cerebral cortex. *J. Neural Eng.* **2005**, *2*, 103–113. [CrossRef] [PubMed]

15. Luan, L.; Wei, X.; Zhao, Z.; Siegel, J.J.; Potnis, O.; Tuppen, C.A.; Lin, S.; Kazmi, S.; Fowler, R.A.; Holloway, S.; et al. Ultraflexible nanoelectronic probes form reliable, glial scar–free neural integration. *Sci. Adv.* **2017**, *3*. [CrossRef] [PubMed]

16. Lo, M.C.; Wang, S.; Singh, S.; Damodaran, V.B.; Kaplan, H.M.; Kohn, J.; Shreiber, D.I.; Zahn, J.D. Coating flexible probes with an ultra fast degrading polymer to aid in tissue insertion. *Biomed. Microdevices* **2015**, *17*, 34–45. [CrossRef]

17. Jeon, M.; Cho, J.; Kim, Y.K.; Jung, D.; Yoon, E.-S.; Shin, S.; Cho, I.-J. Partially flexible MEMS neural probe composed of polyimide and sucrose gel for reducing brain damage during and after implantation. *J. Micromech. Microeng.* **2014**, *24*. [CrossRef]

18. Lecomte, A.; Castagnola, V.; Descamps, E.; Dahan, L.; Blatché, M.C.; Dinis, T.M.; Leclerc, E.; Egles, C.; Bergaud, C. Silk and PEG as means to stiffen a parylene probe for insertion in the brain: Toward a double time-scale tool for local drug delivery. *J. Micromech. Microeng.* **2015**, *25*. [CrossRef]

19. Levental, I.; Georges, P.C.; Janmey, P.A. Soft biological materials and their impact on cell function. *Soft Matter* **2007**, *3*, 299–306. [CrossRef]

20. Gefen, A.; Gefen, N.; Zhu, Q.; Raghupathi, R.; Margulies, S.S. Age-dependent changes in material properties of the brain and braincase of the rat. *J. Neurotrauma* **2003**, *20*, 1163–1177. [CrossRef]

21. Stiller, A.; Usoro, J.; Frewin, C.; Danda, V.; Ecker, M.; Joshi-Imre, A.; Musselman, K.; Voit, W.; Modi, R.; Pancrazio, J.; et al. Chronic intracortical recording and electrochemical stability of Thiol-ene/acrylate shape memory polymer electrode arrays. *Micromachines* **2018**, *9*, 500. [CrossRef]

22. Mather, P.T.; Luo, X.; Rousseau, I.A. Shape memory polymer research. *Annu. Rev. Mater. Res.* **2009**, *39*, 445–471. [CrossRef]

23. Wang, K.; Strandman, S.; Zhu, X.X. A mini review: Shape memory polymers for biomedical applications. *Front. Chem. Sci. Eng.* **2017**, *11*. [CrossRef]

24. Leng, J.; Lan, X.; Liu, Y.; Du, S. Shape-memory polymers and their composites: Stimulus methods and applications. *Prog. Mater. Sci.* **2011**, *56*, 1077–1135. [CrossRef]

25. Black, B.J.; Ecker, M.; Stiller, A.; Rihani, R.; Danda, V.R.; Reed, I.; Voit, W.E.; Pancrazio, J.J. In vitro compatibility testing of thiol-ene/acrylate-based shape memory polymers for use in implantable neural interfaces. *J. Biomed. Mater. Res. Part A* **2018**, *106*, 2891–2898. [CrossRef]

26. Schönfeld, L.M.; Dooley, D.; Jahanshahi, A.; Temel, Y.; Hendrix, S. Evaluating rodent motor functions: Which tests to choose? *Neurosci. Biobehav. Rev.* **2017**, *83*, 298–312. [CrossRef]

27. Schönfeld, L.M.; Jahanshahi, A.; Lemmens, E.; Schipper, S.; Dooley, D.; Joosten, E.; Temel, Y.; Hendrix, S. Long-term motor deficits after controlled cortical impact in rats can be detected by fine motor skill tests but not by automated gait analysis. *J. Neurotrauma* **2017**, *34*, 505–516. [CrossRef]

28. Schneider, C.A.; Rasband, W.S.; Eliceiri, K.W. NIH Image to ImageJ: 25 years of image analysis. *Nat. Methods* **2012**, *9*, 671–675. [CrossRef]

29. Schindelin, J.; Arganda-Carreras, I.; Frise, E.; Kaynig, V.; Longair, M.; Pietzsch, T.; Preibisch, S.; Rueden, C.; Saalfeld, S.; Schmid, B.; et al. Fiji: An open-source platform for biological-image analysis. *Nat. Methods* **2012**, *9*, 676–682. [CrossRef]

30. Stiller, A.; Black, B.; Kung, C.; Ashok, A.; Cogan, S.; Varner, V.; Pancrazio, J. A meta-analysis of intracortical device stiffness and its correlation with histological outcomes. *Micromachines* **2018**, *9*, 443. [CrossRef]
31. Hosseini, S.M.; Rihani, R.; Batchelor, B.; Stiller, A.M.; Pancrazio, J.J.; Voit, W.E.; Ecker, M. Softening shape memory polymer substrates for bioelectronic devices with improved hydrolytic stability. *Front. Mater.* **2018**, *5*. [CrossRef]

Article

Microelectrode Array based Functional Testing of Pancreatic Islet Cells

Ahmad Alassaf [1,2,3], Matthew Ishahak [1,2], Annie Bowles [1,2] and Ashutosh Agarwal [1,2,*]

1   Department of Biomedical Engineering, University of Miami, Coral Gables, FL 33146, USA;
    a.alassaf@umiami.edu (A.A.); m.ishahak@umiami.edu (M.I.); abowles6@gatech.edu (A.B.)
2   DJTMF Biomedical Nanotechnology Institute, University of Miami, Miami, FL 33136, USA
3   Department of Medical Equipment Technology, Majmaah University, Al Majmaah 11952, Saudi Arabia
*   Correspondence: A.agarwal2@miami.edu; Tel.: +305-243-8925

Received: 11 February 2020; Accepted: 15 May 2020; Published: 17 May 2020

Abstract: Electrophysiological techniques to characterize the functionality of islets of Langerhans have been limited to short-term, one-time recordings such as a patch clamp recording. We describe the use of microelectrode arrays (MEAs) to better understand the electrophysiology of dissociated islet cells in response to glucose in a real-time, non-invasive method over prolonged culture periods. Human islets were dissociated into singular cells and seeded onto MEA, which were cultured for up to 7 days. Immunofluorescent imaging revealed that several cellular subtypes of islets; β, δ, and γ cells were present after dissociation. At days 1, 3, 5, and 7 of culture, MEA recordings captured higher electrical activities of islet cells under 16.7 mM glucose (high glucose) than 1.1 mM glucose (low glucose) conditions. The fraction of the plateau phase (FOPP), which is the fraction of time with spiking activity recorded using the MEA, consistently showed distinguishably greater percentages of spiking activity with high glucose compared to the low glucose for all culture days. In parallel, glucose stimulated insulin secretion was measured revealing a diminished insulin response after day 3 of culture. Additionally, MEA spiking profiles were similar to the time course of insulin response when glucose concentration is switched from 1.1 to 16.7 mM. Our analyses suggest that extracellular recordings of dissociated islet cells using MEA is an effective approach to rapidly assess islet functionality, and could supplement standard assays such as glucose stimulate insulin response.

Keywords: islets of Langerhans; insulin secretion; microelectrode array (MEA); glucose stimulated insulin response; electrochemical transduction

1. Introduction

Islets of Langerhans are three-dimensional (3D) multicellular clusters that range from 50 to 500 μm in diameter [1,2]. As the functional unit of the pancreas, islets maintain glucose homeostasis through the interdependent secretion of hormones from α, β, δ, and γ cells [3]. Dysfunction of the insulin-secreting β cells, arising from autoimmune destruction or insulin-resistance, results in the development of diabetes mellitus [4]. To elucidate the pathophysiology of diabetes, new approaches are being employed to study islets at the cellular level.

The electrophysiology of β cells has been investigated using a patch-clamp technique [5–7]. However, the disadvantages to performing the patch clamp technique are the invasiveness to the sample, technical complexity, and limited recording time (hours) [8]. Microelectrode arrays (MEAs), on the other hand, have been employed to collect electrical activities of islets [9–13]. The advantages of MEAs are the non-invasiveness to the sample, ease of execution, and higher duration of recording time (days) [13,14]. Pfeiffer et al. performed extracellular recordings of whole islets using MEA, a glass holding pipette angled at 30°, and a micromanipulator to control the islet location on top of the recording electrode. They were able to show continuous bursts of spikes at high glucose concentration (15 mM) and concluded that the length of the bursts correlated with the amount of insulin released. The fraction of a plateau phase (FOPP), which is the fraction of time with spiking activity recorded using the MEAs was developed as a metric for beta-cell function [9]. Schonecker et al. and Brouwer et al. also used comparable methods on whole islets to confirm that extracellular recordings correlated with intracellular electrical recordings. Phelps et al. reported a new method for culturing dissociated islet cells on glass coverslips, where they were able to perform detailed imaging studies by super-resolution and live cell microscopy. More importantly, cells in the dissociated human and rat islet cell monolayers (α, β, δ, and γ) were in proportions similar to native 3D islets [15].

Herein, we report MEA recordings of dissociated islet cells as an innovative method to capture the islet function while circumventing the limitations of the previously used techniques. Moreover, standard functional tests, i.e., glucose stimulated insulin secretion (GSIS) assays, were performed concomitantly. Our data correlated measurable parameters of electrical activities by the MEA to the functional secretory response of islet cells at the early time points of culture. Moreover, we were able to determine that detection of electrical activities of the islet cells in response to the high glucose stimulation were sustained throughout the long-term culture whereas insulin responses from GSIS were only detectable at the early time points. Together, this evidence supports the utility of MEA for measuring islet function in a highly sensitive, non-invasive, and real-time manner.

2. Materials and Methods

2.1. Islets Dissociation and Culture

Human pancreatic islets were procured from organ donors at the Human Islet Cell Processing Facility at the Diabetes Research Institute (University of Miami, Miller School of Medicine, Miami, FL, USA), under Institutional Review Board (IRB) approval for use of human tissue for research. Human islets are from approved cadaveric organ donors from which at least one other organ has been approved for transplantation. Since the donors are brain dead, the IRB's from the institutions that isolate the islets consider the tissue as "Exempt" from Human Studies Approval. In this study, islets were obtained from two normal non-diabetic donors, a 51 year old male and a 44 year old female, with body mass indices of 29.5 and 32.8 kg/m^2, respectively.

Human collagen IV stock solution (1 mg/mL, Sigma Aldrich, St. Louis, MO, USA) was prepared and diluted to 50 µg/mL into Hanks' Balanced Salt Solution (HBSS) with Ca^{2+}/Mg^{2+} (Life Technologies, Carlsbad, CA, USA). Prior to collagen coating, all MEAs were placed inside a UV ozone cleaner (Jelight) for 8 min in order to sterilize and activate the MEA surface for protein coating. A 100 µL drop of the diluted collagen was then added to each of the UV ozoned MEAs. MEAs were then incubated with collagen IV overnight in 37 °C and washed 3 times with HBSS with Ca^{2+}/Mg^{2+} right before cells seeding.

For dissociation, islets were collected in a conical tube and centrifuged for 2 min at 800 rpm. 400 µL of warmed 0.05% trypsin (Gibco, Waltham, MA, USA) was used for dissociation after three washes with phosphate buffer saline (PBS) were completed. While islets were suspended in trypsin in a cryogenic vial, gentle agitation was applied to the vial in a 37 °C beads bath for 3 min to help with dissociation. After trypsinization for 3 min, 15 mL of neuronal medium was added to deactivate the trypsin and centrifuged for 6 min at 1400 rpm. Approximately 200 cells/mm^2 were added to each collagen IV coated MEA.

Neuronal culture medium was prepared by supplementing minimum essential medium (MEM, Life Technologies, Carlsbad, CA, USA) with 5% fetal bovine serum (FBS), 2% B-27 (50×, Life Technologies), 1% Penicillin-Streptomycin (100×, Life Technologies), 1% HEPES (1 M, Life Technologies), 1% Glutamax (100×, Life Technologies), 1% Na-pyruvate (100 mM, Life Technologies), and final glucose (Life Technologies) concentration of either 1.1 mM glucose (low glucose media) or 16.7 mM glucose (high glucose media).

2.2. Glucose-Stimulated Insulin Secretion (GSIS) Assay

Insulin secretion was assessed by GSIS of dissociated islets under static incubation. Briefly, dissociated islets cultured on each MEA were incubated for one hour at 37 °C in low glucose media (1.1 mM) followed by a one hour incubation in high glucose media (16.7 mM). After each incubation period, a 500 µL sample of media was collected and insulin concentrations were measured using a human insulin enzyme-linked immunosorbent assay (ELISA) kit (Mercodia, Uppsala, Sweden) after diluting the samples 1:500 in deionized water to ensure measurements were within the range of the ELISA kit.

2.3. Electrophysiological Recordings

Electrical activity was recorded from the dissociated islets during GSIS assays using MEA2100 system (Multi Channel Systems MCS GmbH, Reutlingen, Germany). Dissociated islets were cultured on a standard microelectrode array chip (60MEA200/30IR–TI–GR, Multi Channel Systems) that fits inside the MEA2100 system, which was connected with a temperature (37 °C) controller and an interface board that linked the whole system to a PC computer (Multi Channel Systems). Recordings were done on days 1, 3, 5, and 7 post seeding for 15 min using low glucose media first and then high glucose media for each MEA chip. Multichannel experimenter and analyzer programs were used to do on-line recordings and off-line analysis of data, respectively. Electrical signal from each recording was filtered with a high pass filter (200 Hz) and then a low pass filter (4000 Hz), adapted from a previous study [13], and sampled at 25 kHz. A threshold of 10 times the standard deviation of the noise was set to determine a spike. We submit that we were, in fact, more stringent about the threshold compared to some studies that used 5–6 times the standard deviation of the average noise amplitude [13,16]. For spiking profile plots, a bin size of 10 s was used for the total 30 min of recording, where the total number of spikes in each of these 10 s windows was calculated (30 min recording = 180 windows of 10 s) and used to plot the spiking profile.

2.4. Immunofluorescent Staining

Dissociated islets from were cultured for 7 days on 18 mm glass coverslips (Electron Microscopy Sciences). On day 7, each coverslip was incubated for 10 min at room temperature with 4% ice cold paraformaldehyde (PFA) solution after a quick rinse with warm PBS. Three washes with PBS were followed, where each wash was for 5 min with gentle shaking. Next, 0.1% Triton X-100 with gentle shaking was applied for 10 min at room temperature to permeabilize the cells. Three washes with PBS were followed, where each wash was for 5 min with gentle shaking. Using PBS 10% and 1% donkey serum were prepared and used as a blocking buffer. Samples were incubated at room temperature in a dark place for 1 hour after a 200 µL drop of 10% blocking buffer was applied to each sample. Primary antibodies (rabbit anti-insulin antibody, rat anti-somatostatin antibody, and goat anti-pancreatic polypeptide antibody, all purchased from Abcam) solution was prepared as 1:200 dilution with 1% donkey serum and 0.25% Triton-X100. After incubation for 1 hour with the blocking buffer, 200 µL drop of the primary antibodies solution was applied to each sample and then placed inside dark 4 °C fridge overnight.

On the second day, three washes with PBS and 0.01% Triton-X100 were followed, where each wash was for 5 min with gentle shaking. Secondary antibodies (Alexa Fluor 488 donkey anti-rabbit (Life Technologies), Alexa Fluor 555 donkey anti-rat (Abcam), and Alexa Fluor 594 donkey anti-goat (Abcam) solution was prepared as 1:500 dilution in addition to 4′,6-diamidino-2-phenylindole (DAPI, 1:200), where all were mixed with 1% donkey serum and 0.25% Triton-X100. 200 µL drop of the secondary antibodies solution was applied to each sample and incubated for 1 hour at room temperature in a dark place. Coverslips were then rinsed with PBS three times and mounted onto glass slides with ProLong Gold Anti-Fade Reagent (Life Technologies) and sealed with nail polish after curing of the mountant. Stained dissociated islets were imaged on a Nikon Eclipse Ti inverted fluorescent microscope with an Andor Zyla sCMOS camera using a 60× oil immersion objective.

2.5. Statistical Analyses

All statistical analyses were performed on Prism v8 software (GraphPad, San Diego, CA, USA). Paired student *t*-tests were used for statistical comparisons between the low and high glucose conditions for the different days. All values were reported as the mean ± standard error of the mean unless reported otherwise, and $p < 0.05$ was considered statistically significant.

3. Results

3.1. Dissociation and Culture of Islets

Extracellular recordings of intact islets using MEA platform (Figure 1A) and classical MEA chips require proper contact and adhesion between cells and electrodes. Given that islets are large multicellular spheroids, limited contact area with the recording electrodes of planar MEA precludes MEA recording for functional evaluation of islets. Thus, we first enzymatically dissociated islets into single cells (Figure 1B,C) and cultured as adherent cells (hours; Figure 1D) on MEA electrodes to improve cell contacts and accuracy of recorded electrical activity. Fluorescence images on day 7 post seeding on the substrate (Figure 1E) showed that this technique dissociated and retained islet cell types, e.g., β, δ, and γ cells. After dissociation, extracellular recordings were performed the following day (day 1) and until the end culture (day 7).

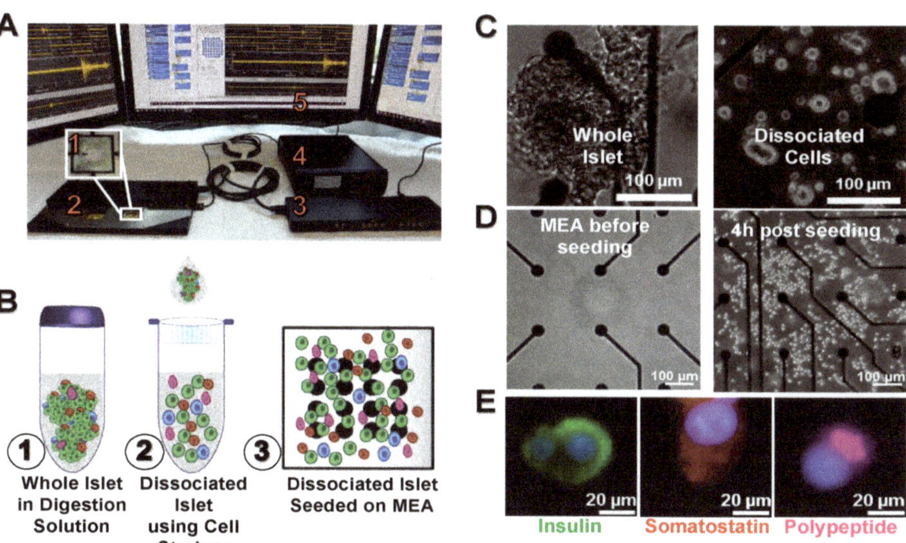

Figure 1. Dissociated islets on a microelectrode array (MEA). (**A**) MEA system setup contains five components; an MEA chip (A1), a two head-stages of MEA2100 system (A2), an interface board (A3), a temperature controller (A4), and a computer (A5). (**B**) Schematic illustration of clusters of whole islets in digestion solution (B1) dissociated into separated islets using a 40 µm cell strainer (B2), and finally seeding these dissociated islets on MEA coated with collagen IV (B3). (**C**) Bright field images of whole islet and dissociated islets on MEA. (**D**) Bright field images of MEA before seeding, and after four hours of seeding the dissociated islets. (**E**) Fluorescent images on day 7 for dissociated islets on glass coverslip showing successful separation of different cell population within the islet. Green color represents insulin (indicating β-cells), red color represents somatostatin (indicating δ-cells), magenta color represents pancreatic polypeptide (indicating γ-cells), and blue color represents DNA (indicating the cell nucleus).

3.2. Extracellular Recordings of Dissociated Islets

Pancreatic β cells show oscillatory electrical activity known as slow waves in response to glucose [5,17]. Extracellular recordings using MEA were previously performed and compared to intracellular measurements obtained by traditional techniques with intact islets to demonstrate an alternate detection method to interrogate the electrical activity of islets [9]. MEA recordings performed using the dissociated islet cells that were seeded on four separate MEA chips (MEA1, MEA2, MEA3, and MEA4). The recordings were for a duration of 30 min on subsequent days, and they showed that high glucose induced longer electrical activity (minutes) with higher amplitudes (mV) compared to the low glucose recordings, which showed shorter electrical activity (seconds) with lower amplitudes (μV; Figure 2). MEA1 on day 3 and MEA2 on day 5, however, showed almost no electrical activity when the high glucose was introduced to the culture. Furthermore, the total number of spikes of the 30-minute recording from each MEA was quantified and the spiking profile plot was generated after binning the number of spikes every 10 s (Figure 3). Interestingly, the time course of binned spike profiles reveal electrophysiological activity that might be associated with the first (immediate) and second (sustained plateau) phase of insulin secretion that is usually seen during dynamic GSIS measurements [18].

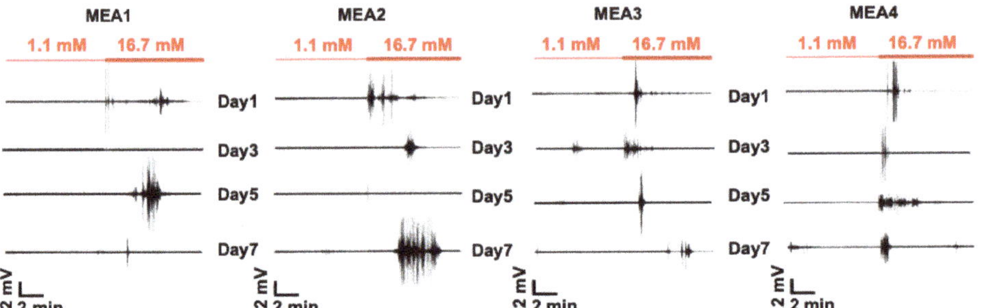

Figure 2. Electrical activity of dissociated human islets. Representative MEA recordings showing electrical activity of islet cells induced by switching from 1.1 to 16.7 mM glucose for different batches of human islets across 7 days.

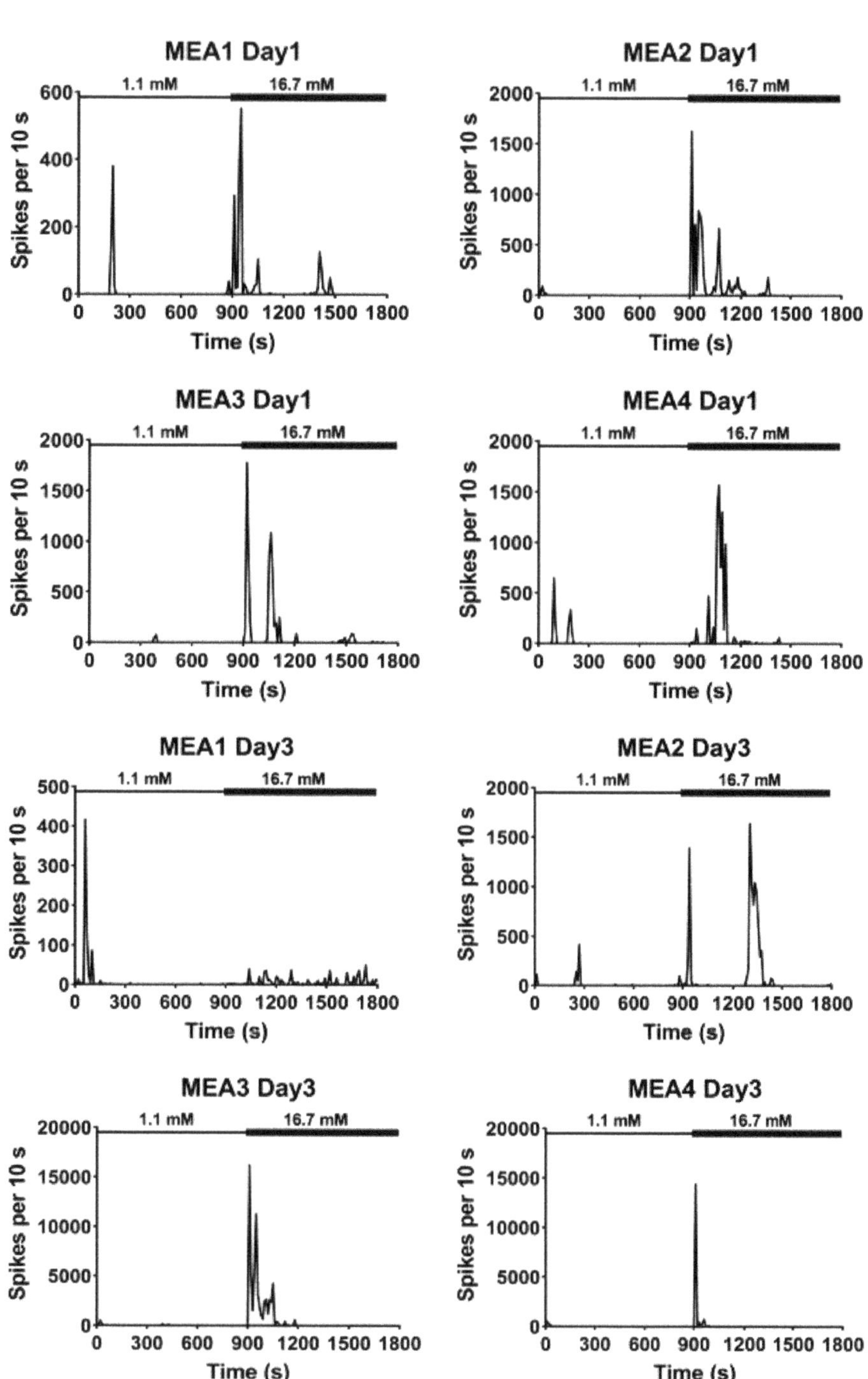

Figure 3. *Cont.*

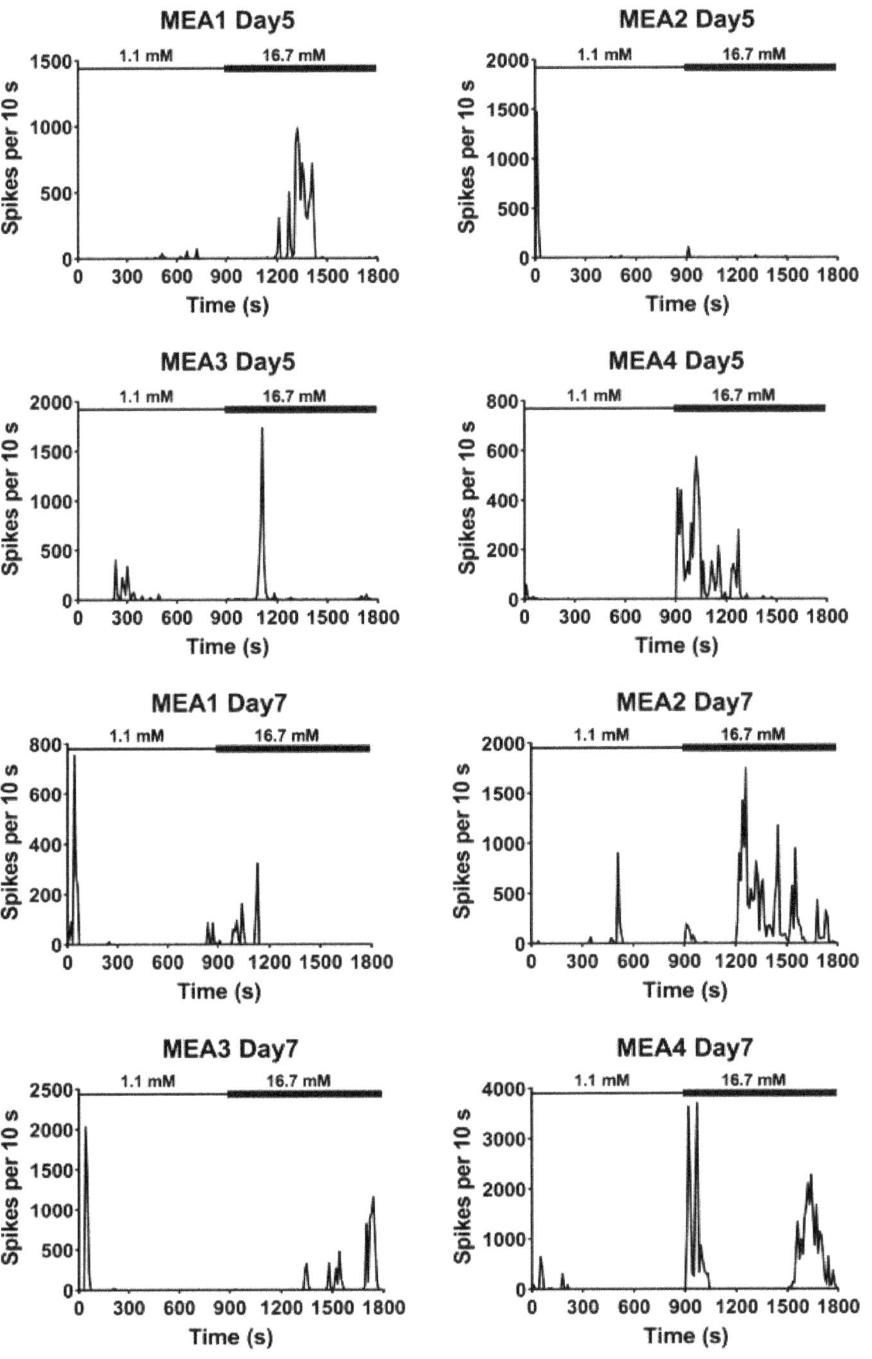

Figure 3. Total spiking with a bin size of 10 s for a total of 30 min recording. Each MEA was recorded with low glucose (1.1 mM) media for 900 s followed by another 900 s recording with high glucose (16.7 mM) media starting from day 1 post seeding until day 7. The number of spikes were summed every 10 s until the end of the recording.

3.3. Insulin Secretion and FOPP

β cells release insulin in response to varying blood glucose levels in vivo, thus measuring insulin concentrations upon exposure to varying levels of glucose in vitro is used as a method to correlate islet function. Standard functional testing using GSIS assays were performed simultaneously to the MEA recordings. Measured insulin concentrations under low (2989 ± 260 µg/L) and high (3275 ± 350 µg/L) glucose at day 1 were comparable to low (2683 ± 95 µg/L) and high (3177 ± 225 µg/L) glucose at day 3 (Figure 4A). However, insulin secretion decreased by day 5 (low G: 3132 ± 284 µg/L, and high G: 2930 ± 147 µg/L) and day 7 (low G: 3178 ± 215 µg/L, and high G: 2618 ± 226 µg/L) of culture. Figure S1 shows the insulin concentrations measurements for each MEA. Interestingly, functional testing using MEA showed FOPP measurements with continuous and high spiking activity with high glucose relative to the low glucose for all culture days. The FOPP calculations were low G: 6% ± 1.3% and high G: 46% ± 4.7% for day 1, low G: 33% ± 5.9% and high G: 40% ± 14.9% for day 3, low G: 17% ± 4.0% and high G: 50% ± 18% for day 5, and low G: 12% ± 1.8% and high G: 44% ± 13.8% for day 7 (Figure 4B). Even though the significant difference between the high and low glucose groups was only on day 1, but the other days showed distinguishable difference. Figure S2 shows the FOPP calculations for each MEA separately.

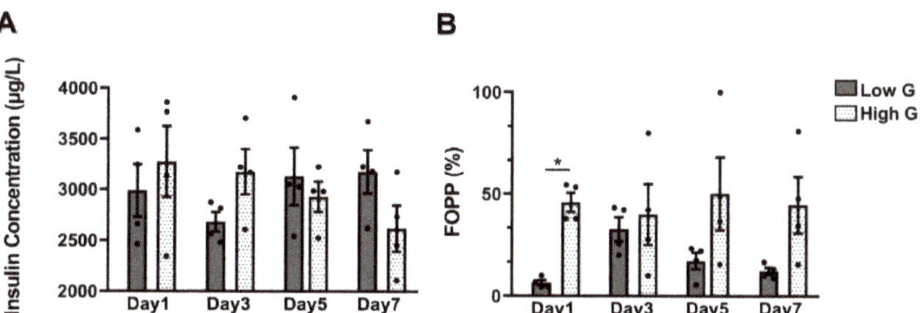

Figure 4. Assessment of insulin secretion and fraction of the plateau phase (FOPP). (**A**) Mean insulin concentration under static incubation of 4 released by dissociated islets on MEAs in low (1.1 mM) and high (16.7 mM) glucose media for different culture days (n = 4 MEAs for each condition). (**B**) Quantification of the mean FOPP measured by MEAs with dissociated islets in response to low (1.1 mM) and high (16.7 mM) glucose media for the different culture days (n = 4 MEAs for each condition). * $p < 0.05$.

4. Discussion

Diabetes mellitus is estimated to affect over 400 million people worldwide by 2030 making it one of the most common and costly chronic diseases [3]. Diabetes is characterized by hyperglycemia related to autoimmune destruction of insulin-secreting β cells (type 1) or insulin resistance (type 2) [19,20]. Islet transplantation is a therapeutic alternative for β-cell replacement, which restores glycemic control in type 1 diabetes patients [21,22]. Islet function can be investigated by traditional assay such as GSIS [18], or by utilizing emerging MEA technology [11,23], which provides information about the islet electrophysiology to test the islet function.

Elucidating the complex and dynamic physiologic processes of healthy islets is imperative prior to transplantation. In general, varying blood glucose levels lead to changes in the membrane potential of β cells inducing an electrochemical mechanism resulting in the release of insulin [12,24–26]. More specifically, increased blood glucose concentration fuels glucose metabolism within β cells, and the product of glycolysis is adenosine triphosphate (ATP). The produced ATP reduces the resting membrane potential, which leads to membrane depolarization (electrical activity). After membrane depolarization, the voltage-gated Ca^{2+} channels open, increasing intracellular Ca^{2+} concentrations, which trigger fusion of vesicles containing insulin with the cell membrane, and subsequent exocytosis of insulin. Insulin is then released into the blood to allow all cells of the body to utilize glucose for energy [5,12].

MEA recordings using human islets were previously measured and electrical activity was detected with high glucose concentration [23,27]. Schonecker et al. showed no electrical activity corresponded to 1 mM glucose, while oscillatory activity (60–80 μV) was evoked by 10 mM glucose concentration when recording from whole islets [23]. In our study, moderate electrical activity (μV) was seen with 1.1 mM glucose, while high spiking (mV) was observed under 16.7 mM glucose concentration. The difference in amplitude between our study and the Schonecker et al. study may be inferred by several factors. Firstly, the Schonecker et al. study used the whole islet, which means only a small portion of the islet surface was in contact with the electrode, whereas we used dissociated islet that had more cells surface area in contact with an electrode. While our seeding density was 50,000 cells on the MEA, the Schonecker et al. study recorded from one whole islet. Additionally, where the Schonecker et al study used only a low pass filter of 100 Hz to filter their signal, compared to our study that used a high pass 200 Hz filter followed by a low pass 4000 Hz filter, which was adapted from Raoux et al' study. Lastly, Schonecker et al. used mouse islets, while our studies exclusively utilized human islet cells.

Insulin exocytosis has been studied and known to follow a biphasic time course [24,28–31]. The first phase linked to a rapid transient increase rate of insulin secretion, commonly within 5 min of glucose stimulation. Then, insulin secretion decreased to a plateau (second phase) before it completely stopped with the end of glucose stimulus. This data suggests that the spiking initiation of this first and second phases in some of the MEAs from the extracellular recording when we plotted the total 30 min spiking profile after we used a 10 s bin size of and calculated the total number of spikes in each MEA. This is the first study to date that correlates spiking profiles to islet function during glucose stimulated insulin secretion of dissociated islets demonstrating an attractive use of MEA.

The length of the spiking when using MEA has been correlated with the amount of insulin released during glucose stimulation [9,12]. Therefore, FOPP assessment for all MEAs on the different days always showed higher percentages of FOPP during the high glucose condition compared to the low glucose. This evidence supports using MEA as a highly sensitive and robust tool to measure the function of human islets. Insulin secretion measurements using the conventional GSIS assay, on the other hand, showed the same trend between high glucose and low glucose conditions only on the early time points of culture. This could suggest the need for cell–cell contact, which was lost in the dissociated islets, and the cooperating cells may require direct intercommunication to secrete higher levels of insulin under high glucose concentrations. Together, these spiking profiles that correspond to glucose stimulated insulin secretion validates the use of MEA for examining islet function.

This study provided supportive evidence that extracellular recordings using MEA is non-invasive and a quick approach that could be used to test islet functionality. By dissociating the islets, the individual islet cells can be cultured, monitored, and recorded for an extended period, which was not possible before MEA technology. Future improvement to this platform includes seeding at various densities to determine the correlation with spiking activity and physiological insulin levels from islets cells. The next logical step to this study and previous MEA studies should be recording from MEA-based in vitro disease models (diabetic islets) and comparing that with the normal healthy islets (baseline) was done in this study. Diabetic islets once obtained, the use of diazoxide, tolbutamide, and KCL could be investigated to show their effect on the electrical activity. Furthermore, our methods can be used to screen new drugs as well as evaluate some of intervention strategies that could be performed when islets are chronically challenged by glucolipotoxicity or stress-inducing agents.

5. Conclusions

In conclusion, we demonstrated that extracellular recordings of dissociated islet cells using MEA is an effective approach to rapidly assess islet functionality, and could supplement standard assays such as glucose stimulate insulin response. Evidence from this study demonstrated islet dissociation, and creating layer of islet cells on planar MEA electrodes is critical for the assay. MEA recordings showed more electrical activity and FOPP percentages induced by the high glucose compared to the low glucose recordings. Furthermore, spiking profile plots from multiple MEA recordings revealed the electrophysiological activity that precedes the initiation of the first (immediate) phase and second (sustained) phase of insulin secretion usually seen during dynamic GSIS measurements. Our approach of dissociating the islets and utilizing the MEA platform to non-invasively test islet functionality is enabling detailed efforts to study islet physiology and screen potential pharmacological interventions

Supplementary Materials: The following are available online at http://www.mdpi.com/2072-666X/11/5/507/s1, Figure S1: Glucose stimulated insulin secretion for functional assessments. Insulin concentration by dissociated islets on each MEA under low (1.1 mM) and high (16.7 mM) glucose media for the different culture days, Figure S2: Assessment of FOPP. Quantification of FOPP of each MEA for low (1.1 mM) and high (16.7 mM) glucose media for the different culture days.

Author Contributions: Conceptualization, A.A. (Ahmad Alassaf) and A.A. (Ashutosh Agarwal); Data curation, A.A. (Ahmad Alassaf) and M.I.; Formal analysis, A.A. (Ahmad Alassaf), M.I. and A.B.; Funding acquisition, A.A. (Ashutosh Agarwal); Investigation, A.A. (Ahmad Alassaf), M.I. and A.B.; Methodology, A.A. (Ahmad Alassaf) and A.A. (Ashutosh Agarwal); Project administration, A.A. (Ashutosh Agarwal); Resources, A.B. and A.A. (Ashutosh Agarwal); Supervision, A.A. (Ashutosh Agarwal); Writing—original draft, A.A. (Ahmad Alassaf) and A.A. (Ashutosh Agarwal); Writing—review & editing, A.A. (Ahmad Alassaf), M.I., A.B. and A.A. (Ashutosh Agarwal). All authors have read and agreed to the published version of the manuscript.

Funding: This work was supported by NIDDK-supported Human Islet Research Network (HIRN, RRID: SCR_014393; https://hirnetwork.org; UC4DK104209 to A.A. and UG3DK122638 to A.A.). M.I. is supported by F31DK118860-01A1.

Conflicts of Interest: The authors declare no conflict of interest.

References

1. Lehmann, R.; Zuellig, R.A.; Kugelmeier, P.; Baenninger, P.B.; Moritz, W.; Perren, A.; Clavien, P.A.; Weber, M.; Spinas, G.A. Superiority of small islets in human islet transplantation. *Diabetes* **2007**, *56*, 594–603. [CrossRef] [PubMed]

2. Elayat, A.A.; el-Naggar, M.M.; Tahir, M. An immunocytochemical and morphometric study of the rat pancreatic islets. *J. Anat.* **1995**, *186 Pt 3*, 629–637.

3. Buse, J.; Polonsky, K.; Burant, C. *Williams Textbook of Endocrinology*, 11th ed.; Elsevier Health Sciences: Amsterdam, The Netherlands, 2011; pp. 1371–1435.

4. Burrack, A.L.; Martinov, T.; Fife, B.T. T Cell-Mediated Beta Cell Destruction: Autoimmunity and Alloimmunity in the Context of Type 1 Diabetes. *Front. Endocrinol. (Lausanne)* **2017**, *8*, 343. [CrossRef] [PubMed]

5. Drews, G.; Krippeit-Drews, P.; Düfer, M. Electrophysiology of islet cells. *Adv. Exp. Med. Biol.* **2010**, *654*, 115–163. [CrossRef] [PubMed]

6. Henquin, J. Regulation of insulin secretion: A matter of phase control and amplitude modulation. *Clin. Exp. Diabetes Metab.* **2009**, *52*, 739–751. [CrossRef] [PubMed]

7. Kanno, T.; Rorsman, P.; Gopel, S.O. Glucose-dependent regulation of rhythmic action potential firing in pancreatic beta-cells by K(ATP)-channel modulation. *J. Physiol.* **2002**, *545*, 501–507. [CrossRef]

8. Shuvaev, A.N.; Salmin, V.V.; Kuvacheva, N.V.; Pozhilenkova, E.A.; Morgun, A.V.; Lopatina, O.L.; Salmina, A.B.; Illarioshkin, S.N. Current advances in cell electrophysiology: Applications for the analysis of intercellular communications within the neurovascular unit. *Rev. Neurosci.* **2016**, *27*, 365. [CrossRef]

9. Pfeiffer, T.; Kraushaar, U.; Düfer, M.; Schonecker, S.; Haspel, D.; Günther, E.; Drews, G.; Krippeit-Drews, P. Rapid functional evaluation of beta-cells by extracellular recording of membrane potential oscillations with microelectrode arrays. *Eur. J. Physiol.* **2011**, *462*, 835–840. [CrossRef]

10. Brouwer, S.; Hoffmeister, T.; Gresch, A.; Schönhoff, L.; Düfer, M. Resveratrol influences pancreatic islets by opposing effects on electrical activity and insulin release. *Mol. Nutr. Food Res.* **2018**, *62*. [CrossRef]

11. Perrier, R.; Pirog, A.; Jaffredo, M.; Gaitan, J.; Catargi, B.; Renaud, S.; Raoux, M.; Lang, J. Bioelectronic organ-based sensor for microfluidic real-time analysis of the demand in insulin. *Biosens. Bioelectron.* **2018**, *117*, 253–259. [CrossRef]

12. Schonecker, S.; Kraushaar, U.; Düfer, M.; Sahr, A.; Hrdtner, C.; Guenther, E.; Walther, R.; Lendeckel, U.; Barthlen, W.; Krippeit-Drews, P.; et al. Long-term culture and functionality of pancreatic islets monitored using microelectrode arrays. *Integr. Biol.* **2014**, *6*, 540–544. [CrossRef] [PubMed]

13. Raoux, M.; Bornat, Y.; Quotb, A.; Catargi, B.; Renaud, S.; Lang, J. Non-invasive long-term and real-time analysis of endocrine cells on micro-electrode arrays. *J. Physiol.* **2012**, *590*, 1085–1091. [CrossRef] [PubMed]
14. Micholt, E.; Jans, D.; Callewaert, G.; Bartic, C.; Lammertyn, J.; Nicolaï, B. Extracellular recordings from rat olfactory epithelium slices using micro electrode arrays. *Sens. Actuators B Chem.* **2013**, *184*, 40–47. [CrossRef]
15. Phelps, E.; Cianciaruso, C.; Santo-Domingo, J.; Pasquier, M.; Galliverti, G.; Piemonti, L.; Berishvili, E.; Burri, O.; Wiederkehr, A.; Hubbell, J.; et al. Advances in pancreatic islet monolayer culture on glass surfaces enable super-resolution microscopy and insights into beta cell ciliogenesis and proliferation. *Sci. Rep.* **2017**, *7*. [CrossRef]
16. Tovar, K.R.; Bridges, D.C.; Wu, B.; Randall, C.; Audouard, M.; Jang, J.; Hansma, P.K.; Kosik, K.S. Action potential propagation recorded from single axonal arbors using multielectrode arrays. *J. Neurophysiol.* **2018**, *120*, 306. [CrossRef]
17. Fridlyand, L.E.; Tamarina, N.; Philipson, L.H. Bursting and calcium oscillations in pancreatic beta-cells: Specific pacemakers for specific mechanisms. *Am. J. Physiol. Endocrinol. Metab.* **2010**, *299*, E517–E532. [CrossRef]
18. Buchwald, P.; Tamayo-Garcia, A.; Manzoli, V.; Tomei, A.A.; Stabler, C.L. Glucose-stimulated insulin release: Parallel perifusion studies of free and hydrogel encapsulated human pancreatic islets. *Biotechnol. Bioeng.* **2018**, *115*, 232–245. [CrossRef]
19. American Diabetes Association. Diagnosis and classification of diabetes mellitus. *Diabetes Care* **2009**, *32* (Suppl. 1), S62–S67. [CrossRef]
20. Aathira, R.; Jain, V. Advances in management of type 1 diabetes mellitus. *World J. Diabetes* **2014**, *5*, 689–696. [CrossRef]
21. Bruni, A.; Gala-Lopez, B.; Pepper, A.R.; Abualhassan, N.S.; Shapiro, A.J. Islet cell transplantation for the treatment of type 1 diabetes: Recent advances and future challenges. *Diabetes Metab. Syndr. Obes.* **2014**, *7*, 211–223. [CrossRef]
22. Correa-Giannella, M.L.; Raposo do Amaral, A.S. Pancreatic islet transplantation. *Diabetol. Metab. Syndr.* **2009**, *1*, 9. [CrossRef] [PubMed]
23. Schonecker, S.; Kraushaar, U.; Günther, E.; Häring, H.-U.; Königsrainer, A.; Drews, G.; Krippeit-Drews, P.; Gerst, F.; Ullrich, S. Human islets exhibit electrical activity on microelectrode arrays (MEA). *Exp. Clin. Endocrinol. Diabetes* **2015**, *123*, 296–298. [CrossRef] [PubMed]
24. Castiello, F.R.; Heileman, K.; Tabrizian, M. Microfluidic perfusion systems for secretion fingerprint analysis of pancreatic islets: Applications, challenges and opportunities. *Lab Chip* **2016**, *16*, 409–431. [CrossRef] [PubMed]
25. Rorsman, P.; Ashcroft, F.M. Pancreatic β-Cell Electrical Activity and Insulin Secretion: Of Mice and Men. *Physiol. Rev.* **2018**, *98*, 117–214. [CrossRef]
26. Fridlyand, L.E.; Jacobson, D.A.; Philipson, L.H. Ion channels and regulation of insulin secretion in human β-cells: A computational systems analysis. *Islets* **2013**, *5*, 1–15. [CrossRef]
27. Udo, K.; Sven, S.; Peter, K.-D.; Gisela, D.; Elke, G. Parallelization of MEA-based electrophysiological recordings of murine and human islets of Langerhans. *Front. Neurosci.* **2016**, *10*. [CrossRef]
28. Lefèbvre, P.; Paolisso, G.; Scheen, A.; Henquin, J. Pulsatility of insulin and glucagon release: Physiological significance and pharmacological implications. *Clin. Exp. Diabetes Metab.* **1987**, *30*, 443–452. [CrossRef]
29. Polonsky, K.S.; Given, B.D.; Van Cauter, E. Twenty-four-hour profiles and pulsatile patterns of insulin secretion in normal and obese subjects. *J. Clin. Investig.* **1988**, *81*, 442–448. [CrossRef]
30. Henquin, J.-C.; Ishiyama, N.; Nenquin, M.; Ravier, M.A.; Jonas, J.-C. Signals and pools underlying biphasic insulin secretion. *Diabetes* **2002**, *51* (Suppl. 1), S60–S67. [CrossRef]
31. Henquin, J.-C.; Nenquin, M.; Stiernet, P.; Ahren, B. In Vivo and In Vitro Glucose-Induced Biphasic Insulin Secretion in the Mouse: Pattern and Role of Cytoplasmic Ca^{2+} and Amplification Signals in β-Cells. *Diabetes* **2006**, *55*, 441–451. [CrossRef]

micromachines

Article

Transparent Microelectrode Arrays Fabricated by Ion Beam Assisted Deposition for Neuronal Cell In Vitro Recordings

Tomi Ryynänen [1],*, Ropafadzo Mzezewa [2], Ella Meriläinen [1], Tanja Hyvärinen [2], Jukka Lekkala [1], Susanna Narkilahti [2] and Pasi Kallio [1]

[1] Micro and Nanosystems Research Group, Faculty of Medicine and Health Technology, Tampere University, 33720 Tampere, Finland; ella.merilainen@tuni.fi (E.M.); jukka.lekkala@tuni.fi (J.L.); pasi.kallio@tuni.fi (P.K.)

[2] Neuro Group, Faculty of Medicine and Health Technology, Tampere University, 33520 Tampere, Finland; ropafadzo.mzezewa@tuni.fi (R.M.); tanja.k.hyvarinen@helsinki.fi (T.H.); susanna.narkilahti@tuni.fi (S.N.)

* Correspondence: tomi.ryynanen@tuni.fi; Tel.: +358-294-5211

Received: 31 March 2020; Accepted: 12 May 2020; Published: 14 May 2020

Abstract: Microelectrode array (MEA) is a tool used for recording bioelectric signals from electrically active cells in vitro. In this paper, ion beam assisted electron beam deposition (IBAD) has been used for depositing indium tin oxide (ITO) and titanium nitride (TiN) thin films which are applied as transparent track and electrode materials in MEAs. In the first version, both tracks and electrodes were made of ITO to guarantee full transparency and thus optimal imaging capability. In the second version, very thin (20 nm) ITO electrodes were coated with a thin (40 nm) TiN layer to decrease the impedance of Ø30 μm electrodes to one third (1200 kΩ → 320 kΩ) while maintaining (partial) transparency. The third version was also composed of transparent ITO tracks, but the measurement properties were optimized by using thick (200 nm) opaque TiN electrodes. In addition to the impedance, the optical transmission and electric noise levels of all three versions were characterized and the functionality of the MEAs was successfully demonstrated using human pluripotent stem cell-derived neuronal cells. To understand more thoroughly the factors contributing to the impedance, MEAs with higher IBAD ITO thickness as well as commercial sputter-deposited and highly conductive ITO were fabricated for comparison. Even if the sheet-resistance of our IBAD ITO thin films is very high compared to the sputtered one, the impedances of the MEAs of each ITO grade were found to be practically equal (e.g., 300–370 kΩ for Ø30 μm electrodes with 40 nm TiN coating). This implies that the increased resistance of the tracks, either caused by lower thickness or lower conductivity, has hardly any contribution to the impedance of the MEA electrodes. The impedance is almost completely defined by the double-layer interface between the electrode top layer and the medium including cells.

Keywords: microelectrode array (MEA); ion beam assisted electron beam deposition (IBAD); indium tin oxide (ITO); titanium nitride (TiN); neurons; transparent

1. Introduction

Microelectrode or multielectrode arrays (MEAs) are a common measurement platform in various in vitro studies where, for example, neuronal cells or cardiomyocytes are applied for drug screening, toxicity testing, cell model development or simply for increasing understanding of cell behavior [1–4]. The field potential [5] or impedimetric [6] measurements typically performed with MEA are usually complemented with fluorescence imaging or microscopic inspection during or after the MEA recordings. In such studies, the use of an inverted microscope is preferred as imaging from the top side is often impossible because of the cell culturing medium and its reservoir placed on top of the MEA. With the inverted microscope, there exists, however, another challenge. Typically, the tracks and especially the electrodes of the MEA are opaque and thus they prevent the full visibility of the cells from the bottom side—just where the electrical signal is originating and imaging the cells is usually of the greatest interest.

A common and unfortunately only a partial solution to the above problem is to make the tracks from transparent indium tin oxide (ITO) [7–9] material. However, ITO is rarely used as the electrode material [10,11], simply because of the relatively high impedance and noise level of the ITO electrodes. Thus, opaque low-impedance platinum black [12] or titanium nitride (TiN) [13] electrodes are usually used with ITO tracks. Recently both Ryynänen et al. [14] and Mierzejewski et al. [15] have proposed the use of a very thin TiN layer on the electrodes, which would benefit from TiN's columnar structure and thus capability of decreasing impedance by increasing the effective surface area, while still maintaining the transparency, at least to some extent. Graphene [16,17], diamond [18] and conducting polymers [19,20] have also been demonstrated as potential candidates for transparent electrodes, but they have their own challenges especially related to the ease of fabrication and stability.

Sputter deposition is the standard method for depositing both ITO [21,22] and TiN [13] thin films. However, we have successfully shown that ion beam assisted electron beam deposition (IBAD) can also be used as a deposition method in fabrication of opaque TiN electrodes [23]. In this paper, IBAD TiN has been applied not only as an opaque layer, but also as a very thin transparent top layer for the ITO electrodes. In addition, a process for depositing ITO by IBAD was developed. Apart from showing that IBAD is a valid alternative for sputtering in depositing both ITO and TiN for MEAs, this paper also focuses on evaluating the performance of different combinations of ITO tracks, as well as ITO and TiN electrodes, both from imaging and impedance points of view. Biocompatibility of the MEAs was verified by performing cell experiments with human pluripotent stem cell-derived (hPSC) neuronal cells.

2. Materials and Methods

2.1. MEA Design and Fabrication

A total of four batches of MEAs (Table 1) were fabricated and each batch included three versions of MEAs (Table 2). In batch 1 two MEAs of each version were fabricated and in the other batches one MEA of each version. The vertical structure of the MEAs is illustrated in Figure 1a. In version 1, both tracks and electrodes were patterned on an ITO layer. To decrease the impedance and the noise level of the electrodes, the ITO electrodes were coated with a 40 nm TiN layer in version 2 or with a 200 nm TiN layer in version 3. The thin film thicknesses presented in this paper are nominal values measured by the quartz crystal-based thickness controller of the deposition system. Stylus profilometer (Bruker DektakXT) measurements were occasionally performed to check that the thickness was within ~±10% limits. TiN was also applied to the contact pads in each MEA version to increase their thickness and thus make them less prone to be punched by the contact pins. A custom layout included 72 circular measurement electrodes with typical 30 μm diameter and 180–200 μm center-to-center distance. There were also 24 square measurement electrodes with 35×50 μm^2 area and 180 μm center-to-center distance and a large grounding electrode shared between three areas containing an equal amount of both measurement electrode types. The MEAs were compatible with the Multi-Channel Systems' (MCS, Multi-Channel Systems MCS GmbH) 120-electrode MEA format.

Table 1. Details of microelectrode array (MEA) batches.

MEA Batch	1	2	3	4
ITO thickness (nm)	20	20	150	180
ITO deposition method	IBAD	IBAD	IBAD	Sputtering

Table 2. Details of MEA versions.

MEA Version	1	2	3
Track material	ITO	ITO	ITO
Electrode material	ITO	TiN on ITO	TiN on ITO
TiN thickness (nm)	-	40	200

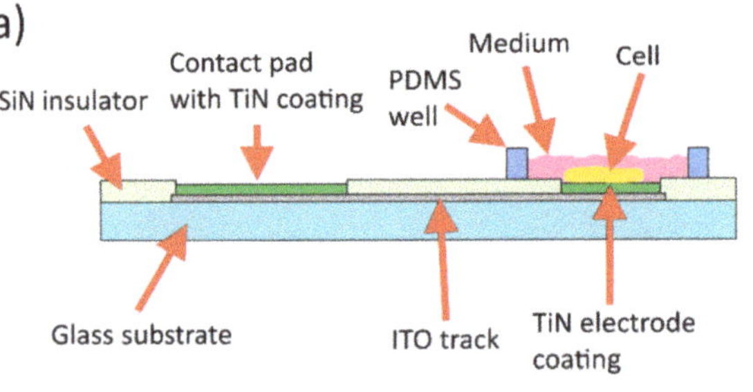

Figure 1. (**a**) Sideview of the ion beam assisted electron beam deposited (IBAD) indium tin oxide (ITO) MEA structure. The image is not to scale. The structure is the same in each batch and version except that the thickness of the ITO track (grey) and/or its deposition method are varied in different batches and the existence or thickness of the TiN coating (green) is varied in different versions. See Tables 1 and 2 for more details. (**b**) Photograph of an IBAD ITO MEA with a polydimethylsiloxane (PDMS) microfluidic device. The dimensions of the glass substrate are 49 mm × 49 mm.

The MEA fabrication process was started by cleaning 49 mm × 49 mm × 1 mm soda lime glass wafers (Gerhard Menzel) with acetone and isopropanol in an ultrasound bath. The ITO layer was deposited on the wafers by the IBAD method, where 95/5 ITO pellets (g-materials) were used as source material and evaporated by e-beam at ~0.24 Å/s deposition rate. During the deposition, the wafers were bombarded with oxygen and argon ions from Saintech ST-55 ion source (Telemark). The gas flow for both gases was 5 sccm (standard cubic centimeters per minute) and the ion current was ~1.1 μA. The fabrication was continued by a lift-off process for titanium alignment marks. After that, electrodes, tracks and contact pads were patterned to the ITO layer by reactive ion etching (RIE) using argon as etching gas and positive photoresist (Futurrex PR1-2000A1) as etching mask. An insulator layer of 500 nm of silicon nitride (SiN) was then deposited at 300 °C by plasma enhanced chemical vapor deposition (PECVD). Positive photoresist was used as the etching mask in etching the openings for electrodes and contact pads by a RIE (Advanced Vacuum Vision 320) process using SF_6 and O_2 as the etching gases. The fabrication of the different versions of the MEAs was finalized by a lift-off process of IBAD deposited TiN as described earlier in [23]. In the case of version 1, the electrodes were protected by a drop of photoresist prior to TiN deposition and thus in version 1 only the contact pads have a TiN layer.

In the literature, the track thickness in MEAs is, typically, a couple of hundred nm [23–25]. However, in this study, in batches 1 and 2, the IBAD ITO thickness was only 20 nm. The motivation for this was the rarely-reported, but commonly-known increased leakage risk of silicone tunnel structures when crossing over the tracks. Using thinner tracks and thus a flatter MEA surface might help to reduce the risk. For comparison, especially to study the role of the track thickness and conductivity on track impedance, two more batches of MEAs were fabricated. The electrode top layers remained the same in each batch, while the underlying track layer was modified. In batch 3, the IBAD layer thickness was increased to 150 nm and in batch 4, commercial 0.7 mm thick boro-aluminosilicate glass wafers with sputter-deposited ITO (University Wafer Inc.) were used as the starting point. In these

MEAs, the ITO thickness was 180 nm and the sheet resistance was as low as 8–10 Ω/sq, according to the manufacturer.

2.2. Technical Characterization

Transmission spectra in the visible wavelength range (380–800 nm) of the 20 nm ITO layer as well as the ITO layer + the thin 40 nm TiN layer were measured with JAZ spectrometer (Ocean Optics). Transmission of the thicker 200 nm TiN layer was not measured because of its obvious opaqueness. Both samples were randomly selected from different deposition batches than those used for fabricating the MEAs.

The impedances of all the 72 + 24 measurement electrodes of each MEA were measured at 1 kHz frequency with MEA-IT120 impedance test device (MCS). In the case of batch 1, which was also the batch used in the cell experiments, the impedances were measured before the cell experiments. Dulbecco's phosphate-buffered saline (DPBS, Merc) was used as the medium in the impedance measurements. The results are presented as a mean ± standard deviation of each functional electrode of the same MEA type and electrode size. Electrodes whose impedance was greater than twice the median impedance or less than half of the median impedance were excluded from the calculations as faulty electrodes. Area-normalized impedance values were calculated by assuming that the impedance is inversely proportional to the electrode area.

The noise levels of each MEA version of batch 1 were approximated by calculating the root-mean-square values from the 10 min cell recording data on day 10 of the cells on MEA. Conductivities of IBAD ITO thin films were evaluated by measuring sheet resistances from test samples. A simple millimeter-sized pattern for 4-point measurement (Figure 2) was deposited through a mechanical mask on a microscope slide and the measurement was done by a Keithley 2002 digital multimeter.

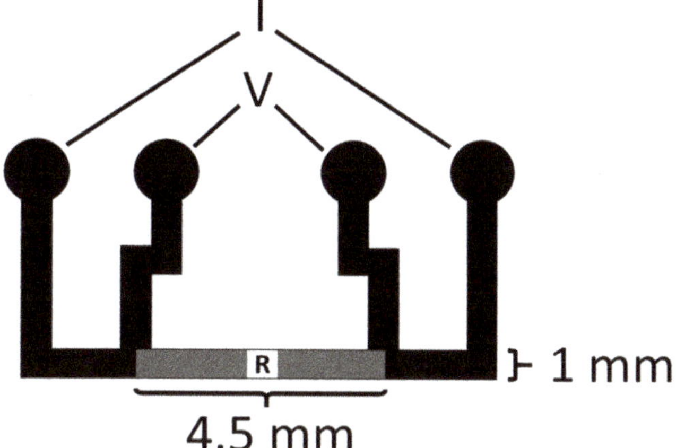

Figure 2. Four-wire test pattern for sheet resistance measurements.

2.3. Cell Experiments

A polydimethylsiloxane (PDMS) microfluidic device with three cell culture medium compartments was bonded onto the MEAs before plating the cells (Figure 1b). A human embryonic stem cell (hESC) line Regea 08/023 for neuronal cell production was used in this study. The Faculty of Medicine and Health Technology has the approval from the Finnish Medicines Agency (FIMEA) to perform research using human embryos (Dnro 1426/32/300/05). The cortical neural differentiation was performed by the recently optimized method [26]. The neuronal cell plating to the devices was performed after 32 days of predifferentiation. This day is considered as day zero on the devices [26]. The cells were plated in a density of 290,000/cm^2 to each compartment. All medium in the medium reservoirs (200 μl/compartment) was changed a day before MEA recordings and in total four times per week. A control cell culture was grown at the same time on standard plastic plates.

The neuronal activity in the devices was recorded using a MEA2100 system (MCS). The temperature of the MEA headstage was kept at a steady +37 °C using TC02 temperature controller (MCS). The medium compartments were covered with a sterile transparent plastic film to prevent evaporation of medium and contamination during the measurements. The standard 10 min recordings for monitoring the development of the network activity were performed twice a week, up to seven weeks (84 days in vitro, 52 days on MEA). The raw data was obtained at the sampling rate of 25 kHz.

3. Results

The transmission spectrum (Figure 3) showed an almost linear increase in transmission from 74% to 89% for the 20 nm IBAD ITO layer within the studied wavelength range. When a 40 nm IBAD TiN layer was added, the transmission dropped to 17%–24%. Importantly, partial transparency was still maintained unlike with the thicker 200 nm IBAD TiN layer, whose transmission was not measured for its obvious opaqueness.

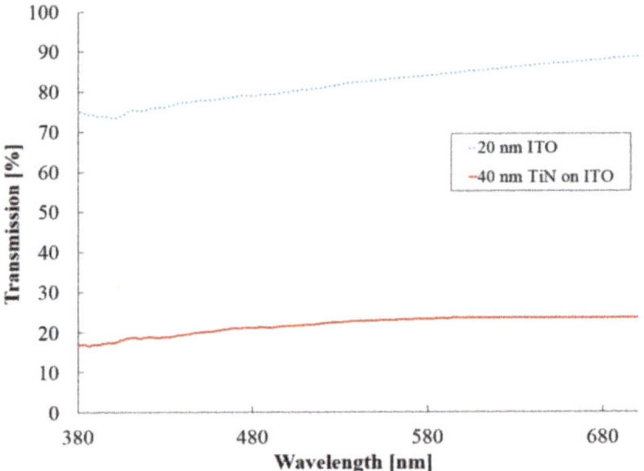

Figure 3. Transmission of IBAD deposited 20 nm ITO and 40 nm TiN films at visible wavelengths.

Impedance measurement results as well as noise level and sheet resistance estimates are summarized in Table 3. Briefly, version 1 with bare ITO electrodes clearly showed the highest impedance, from 1200 kΩ to 1950 kΩ depending on the batch. The errors in these MEAs were also relatively high. In version 2, the thin TiN layer resulted in a drop of the impedance to one quarter or even more when compared to the bare ITO electrodes, ranging between 300 kΩ to 370 kΩ. The thick TiN layer in version 3 resulted in a decrease of the impedances to ~200 kΩ. In square electrodes, possessing a larger area, the impedance values were lower as expected, roughly 50% lower compared with the circular Ø30 μm electrodes. Depending on the MEA, 4–13 (5%–18%) of the 72 circular Ø30 μm electrodes were considered faulty and excluded from the results. The fault was usually caused by a scratch, dirt, or failed lift-off type of fabrication damage while no correlation to ITO thickness nor its source were observed.

Table 3. Impedances at 1 kHz and estimates of RMS noise level and sheet resistance. In the normalized column the impedances of $35 \times 50\ \mu m^2$ electrodes are normalized to correspond to the area of $\varnothing 30\ \mu m$ electrodes.

Batch/ Version	ITO Deposition Method	ITO Thickness (nm)	TiN Thickness (nm)	Measured Impedance of $\varnothing 30\ \mu m$ Electrodes (kΩ)	Measured Impedance of $35 \times 50\ \mu m^2$ Electrodes (kΩ)	Normalized Impedance (kΩ)	RMS Noise (μV)	ITO Sheet Resistance ($\Omega/\square$)
1/1	IBAD	20	-	1200 ± 260	820 ± 140	2030	12 ± 1	2.6×10^3
2/1	IBAD	20	-	1950 ± 810	1170 ± 120	2900	Na	9.4×10^6
3/1	IBAD	150	-	1940 ± 290	1180 ± 150	2920	Na	76.4×10^3
4/1	Sputtering	180	-	1420 ± 170	830 ± 30	2050	Na	8–10 *
2/1	IBAD	20	40	320 ± 20	190 ± 10	470	6.1 ± 0.5	2.6×10^3
2/2	IBAD	20	40	370 ± 10	150 ± 10	370	na	9.4×10^6
3/2	IBAD	150	40	350 ± 10	160 ± 3	400	na	76.4×10^3
4/2	Sputtering	180	40	300 ± 10	130 ± 2	320	na	8–10 *
1/3	IBAD	20	200	190 ± 10	150 ± 20	370	5.6 ± 0.5	2.6×10^3
2/3	IBAD	20	200	240 ± 10	120 ± 2	300	na	9.4×10^6
3/3	IBAD	150	200	210 ± 4	110 ± 2	270	na	76.4×10^3
4/3	Sputtering	180	200	190 ± 3	90 ± 1	220	na	8–10 *

* Data from supplier (University Wafer Inc.).

In the cell experiments, no biocompatibility issues were observed and the neurons grew well during the seven weeks of culture time. Microscopic imaging with the inverted microscope was also successful. With the inverted microscope, it was possible to observe the cells through the electrodes in versions 1 and 2 (Figure 4a,b). In version 3, that was not possible. However, due to the transparent tracks, only the cells located directly on top of the electrodes were invisible (Figure 4c). On the control plastic plate (Figure 4d) the cell density after 29 days was a bit higher than on the MEAs, which was expected, as the cells are known to favor plastic vs. SiN/glass MEA surface. The field potentials with all three MEA versions were successfully recorded (Figure 5). In version 1, the signal peaks were clearly lower compared to the other two versions, which provided signal-to-noise ratio comparable to what is typically seen both with commercial or in-house made MEAs with Ti tracks and TiN electrodes [23].

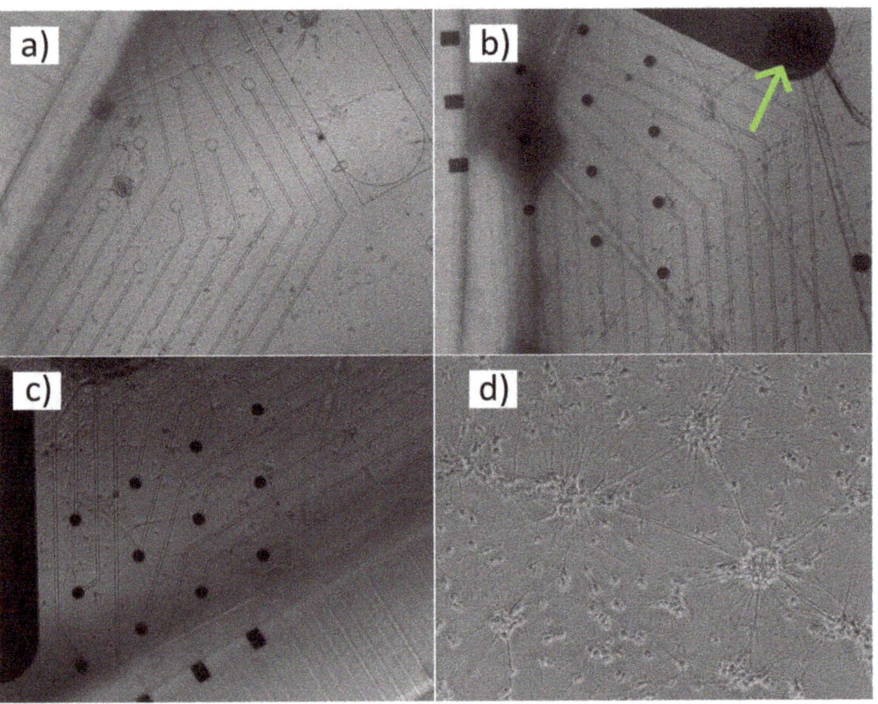

Figure 4. Inverted microscope images of neuronal cell networks growing on different versions of batch 1 MEAs (**a–c**) and a control culture (**d**) grown at the same time on standard plastic plate (29 days after plating). (**a**) The cells are fully visible through the tracks and circular electrodes of MEA version 1. (**b**) In MEA version 2 the visibility through the electrodes drops to partial. However, the cells above the electrodes can still be seen (see especially the larger cluster on the grounding electrode, pointed out by the arrow). (**c**) No visibility through the electrodes in MEA version 3. The control image was taken with a different microscope than the MEA images.

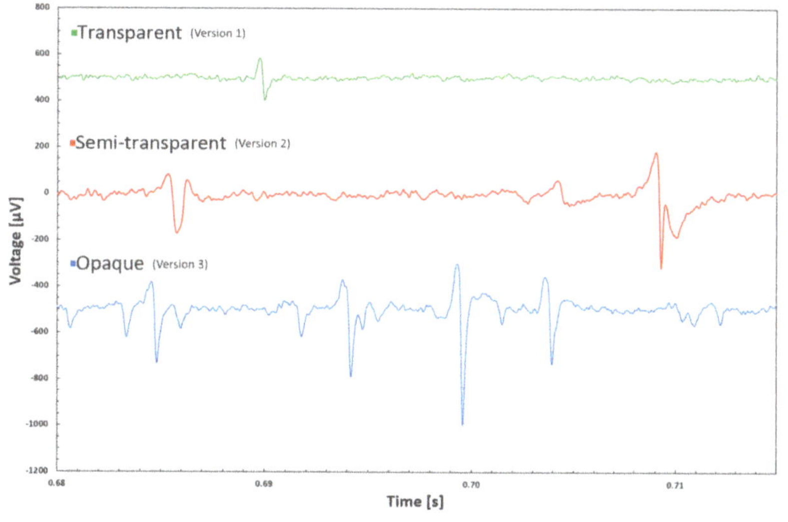

Figure 5. Examples of neuronal cell field potential recordings with different MEA versions from batch 1. Offsets of the curves have been shifted for clarity.

4. Discussion

The fabrication of MEAs with ITO or TiN electrodes has traditionally required access to a sputter coater with corresponding targets and preferably the possibility to run a reactive sputtering process with oxygen and nitrogen gases. In this paper, we have shown that ITO electrodes can be fabricated using an e-beam coater equipped with an ion-source, i.e., by the IBAD method. An alternative for sputtering thus exists.

For opaque TiN electrodes, we have shown the usability of the IBAD method in an earlier paper [23]. Similarly, a decrease in impedance while maintaining (partial) transparency by using a thin TiN layer has been shown earlier to be doable by sputtering [15] and atomic layer deposition (ALD) [14]. In this paper, we show for the first time that transparency of electrodes with reasonable impedance can also be achieved with the IBAD method. Even if the main criteria for the method selection is the availability of the deposition systems, IBAD provides some advantages in terms of raw material costs as one can get started with just two crucibles of pellets compared with large and expensive sputtering targets or the usually not-so-cheap ALD precursors. Furthermore, if a TiN layer is applied, as an e-beam deposition-based line-of-sight method IBAD has better lift-off compatibility than more conformal sputtering [27] and ALD [28] deposition methods. The downside of IBAD is its compromised thickness control compared especially with ALD, in which the thin film thickness can be controlled at atomic layer level [28], at least in theory.

The rather large differences in the impedance values between batch 1 and batch 2, despite their equal parameters most likely originated from the unavoidable differences in the IBAD deposition conditions—our deposition system is manually operated and, in addition, it is also used for many other thin film depositions which typically affect the following depositions. Batches 2–4 were fabricated within a short time and their deposition conditions should be more comparable than the ones of batch 1, fabricated one year earlier. However, for the high error in the impedances of version 1 MEAs, i.e., the MEAs without TiN coating and thus with high impedance (1200–1950 kΩ), the most evident explanation is the MEA-IT impedance tester device. We have noticed that it has difficulty in measuring high impedances in a reliable and repeatable manner as it is optimized for 1–2 decades lower impedances.

When the impedances of square electrodes were normalized to correspond to the area of smaller circular electrodes (Table 3), excluding one lucky exception, there was no match, but the normalized values were higher than the circular electrode values. This is, however, to be expected as the idealistic assumption of the impedance being exactly inversely proportional to the area is rarely true. Usually the impedance has not only resistive but also capacitive components. Another factor may be that the calculations were made based on nominal electrode sizes and it is possible that in different MEAs and electrode shapes the error to the real sizes varies slightly.

In many electronic devices, conductivity of the tracks or wires has a great influence on the operation and performance of the device. In MEAs, however, the impedance of the double-layer interface [29] between the electrode surface and the medium/cells is so huge compared with the rest of the system that the tracks have practically no influence on the total impedance. This was also proved in this study as neither the ITO track deposition method nor thickness made any difference to the impedance which would not fit within the electrode-to-electrode or batch-to-batch variations. On the contrary, adding a TiN coating to the electrodes and increasing its thickness clearly decreased the impedance. Even if not visible by the simplified and, unfortunately, common industry-standard method applied in this study, e.g., evaluating the magnitude of the electrode impedance at 1 kHz frequency, one may miss more subtle differences. Thus, most likely with more detailed noise and impedance analysis over a wider frequency range, evidence may arise that the track conductivity does matter when the ultimate signal quality, especially at certain frequency ranges, is considered. This is, however, left for future studies, partly because our IBAD ITO process is not yet able to produce as highly conductive thin films as has been reported earlier for other applications [30]. So, another future topic would be improving the conductivity of IBAD ITO thin films.

5. Conclusions

To deposit ITO and TiN for MEA tracks and electrodes, IBAD is a valid alternative to sputtering. Depending on the need, the transmission and the impedance of the ITO, electrodes can be tuned by varying the thickness of the TiN layer. A 40 nm TiN layer on ITO electrodes can be considered as a good compromise, providing sufficient transmission in inverted microscope imaging but also relatively low impedance and thus a good signal-to-noise ratio.

Author Contributions: Conceptualization, T.R.; methodology, T.R., E.M., and T.H.; formal analysis, T.R., R.M. and E.M.; investigation, T.R., R.M. and E.M.; writing—original draft preparation, T.R. and R.M.; writing—review and editing, T.R., R.M., T.H., P.K., J.L. and S.N.; visualization, T.R. and R.M.; project administration, P.K., J.L. and S.N.; funding acquisition, P.K., J.L. and S.N. All authors have read and agreed to the published version of the manuscript.

Funding: This research was funded by the Academy of Finland (AF) Centre of Excellence in Body-on-Chip Research (CoEBoC) and AF MEMO project and by the Human Spare Parts project of Business Finland.

Acknowledgments: The authors acknowledge the Tampere facility of Electrophysiological Measurements for their service and Matias Jokinen and Lassi Sukki for providing the PDMS microfluidic devices.

Conflicts of Interest: The authors declare no conflict of interest.

References

1. Stett, A.; Egert, U.; Guenther, E.; Hofmann, F.; Meyer, T.; Nisch, W.; Haemmerle, H. Biological application of microelectrode arrays in drug discovery and basic research. *Anal. Bioanal. Chem.* **2003**, *377*, 486–495. [CrossRef] [PubMed]
2. Defranchi, E.; Novellino, A.; Whelan, M.; Vogel, S.; Ramirez, T.; Van Ravenzwaay, B.; Landsiedel, R. Feasibility Assessment of Micro-Electrode Chip Assay as a Method of Detecting Neurotoxicity in vitro. *Front. Neuroeng.* **2011**, *4*, 6. [CrossRef] [PubMed]
3. Lahti, A.L.; Kujala, V.J.; Chapman, H.; Koivisto, A.-P.; Pekkanen-Mattila, M.; Kerkela, E.; Hyttinen, J.; Kontula, K.; Swan, H.; Conklin, B.R.; et al. Model for long QT syndrome type 2 using human iPS cells demonstrates arrhythmogenic characteristics in cell culture. *Dis. Model. Mech.* **2012**, *5*, 220–230. [CrossRef]

4. Ylä-Outinen, L.; Heikkilä, J.; Skottman, H.; Suuronen, R.; Aänismaa, R.; Narkilahti, S. Human cell-based micro electrode array platform for studying neurotoxicity. *Front. Neuroeng.* **2010**, *3*, 1–9. [CrossRef]

5. Zhu, H.; Scharnhorst, K.S.; Stieg, A.Z.; Gimzewski, J.K.; Minami, I.; Nakatsuji, N.; Nakano, H.; Nakano, A. Two dimensional electrophysiological characterization of human pluripotent stem cell-derived cardiomyocyte system. *Sci. Rep.* **2017**, *7*, 43210. [CrossRef] [PubMed]

6. Eichler, M.; Jahnke, H.G.; Krinke, D.; Müller, A.; Schmidt, S.; Azendorf, R.; Robitzki, A.A. A novel 96-well multielectrode array based impedimetric monitoring platform for comparative drug efficacy analysis on 2D and 3D brain tumor cultures. *Biosens. Bioelectron.* **2015**, *67*, 582–589. [CrossRef]

7. Hammack, A.; Rihani, R.T.; Black, B.J.; Pancrazio, J.J.; Gnade, B.E. A patterned polystyrene-based microelectrode array for in vitro neuronal recordings. *Biomed. Microdevices* **2018**, *20*, 48. [CrossRef]

8. Tang, R.Y.; Pei, W.H.; Chen, S.Y.; Zhao, H.; Chen, Y.F.; Han, Y.; Wang, C.L.; Chen, H.D. Fabrication of strongly adherent platinum black coatings on microelectrodes array. *Sci. China Inf. Sci.* **2014**, *57*, 1–10. [CrossRef]

9. Gross, G.W.; Wen, W.Y.; Lin, J.W. Transparent indium-tin oxide electrode patterns for extracellular, multisite recording in neuronal cultures. *J. Neurosci. Methods* **1985**, *15*, 243–252. [CrossRef]

10. Kim, Y.H.; Kim, G.H.; Baek, N.S.; Han, Y.H.; Kim, A.-Y.; Chung, M.-A.; Jung, S.-D. Fabrication of multi-electrode array platforms for neuronal interfacing with bi-layer lift-off resist sputter deposition. *J. Micromech. Microeng.* **2013**, *23*, 097001. [CrossRef]

11. Nam, Y.; Musick, K.; Wheeler, B.C. Application of a PDMS microstencil as a replaceable insulator toward a single-use planar microelectrode array. *Biomed. Microdevices* **2006**, *8*, 375–381. [CrossRef] [PubMed]

12. Oka, H.; Shimono, K.; Ogawa, R.; Sugihara, H.; Taketani, M. A new planar multielectrode array for extracellular recording: Application to hippocampal acute slice. *J. Neurosci. Methods* **1999**, *93*, 61–67. [CrossRef]

13. Janders, M.; Egert, U.; Stelzle, M.; Nisch, W. Novel thin film titanium nitride micro-electrodes with excellent charge transfer capability for cell stimulation and sensing applications. In Proceedings of the 18th Annual International Conference of the IEEE Engineering in Medicine and Biology Society (1996), Amsterdam, The Netherlands, 31 October–3 November 1996; pp. 245–247.

14. Ryynänen, T.; Pelkonen, A.; Grigoras, K.; Ylivaara, O.M.E.; Hyvärinen, T.; Ahopelto, J.; Prunnila, M.; Narkilahti, S.; Lekkala, J. Microelectrode Array With Transparent ALD TiN Electrodes. *Front. Neurosci.* **2019**, *13*, 226. [CrossRef] [PubMed]

15. Mierzejewski, M.; Kshirsagar, P.; Kraushaar, U.; Heusel, G.; Samba, R.; Jones, P.D. Bringing transparent microelectrodes to market: Evaluation for production and real-world applications. In Proceedings of the Front. Cell. Neurosci. Conference Abstract: MEA Meeting 2018 | 11th International Meeting on Substrate Integrated Microelectrode Arrays, Reutlingen, Germany, 4–6 July 2018.

16. Kireev, D.; Seyock, S.; Lewen, J.; Maybeck, V.; Wolfrum, B.; Offenhäusser, A. Graphene Multielectrode Arrays as a Versatile Tool for Extracellular Measurements. *Adv. Healthc. Mater.* **2017**, *6*, 1601433. [CrossRef] [PubMed]

17. Koerbitzer, B.; Krauss, P.; Nick, C.; Yadav, S.; Schneider, J.J.; Thielemann, C. Graphene electrodes for stimulation of neuronal cells. *2D Mater.* **2016**, *3*, 024004. [CrossRef]

18. Granado, T.C.; Neusser, G.; Kranz, C.; Filho, J.B.D.; Carabelli, V.; Carbone, E.; Pasquarelli, A. Progress in transparent diamond microelectrode arrays. *Phys. Status Solidi A* **2015**, *212*, 2445–2453. [CrossRef]

19. Kshirsagar, P.; Burkhardt, C.; Mierzejewski, M.; Chassé, T.; Fleischer, M.; Jones, P.D. Graphene-based transparent microelectrode arrays for optical access to the recording site. In Proceedings of the Front. Cell. Neurosci. Conference Abstract: MEA Meeting 2018 | 11th International Meeting on Substrate Integrated Microelectrode Arrays, Reutlingen, Germany, 4–6 July 2018.

20. Blau, A.; Murr, A.; Wolff, S.; Sernagor, E.; Medini, P.; Iurilli, G.; Ziegler, C.; Benfenati, F. Flexible, all-polymer microelectrode arrays for the capture of cardiac and neuronal signals. *Biomaterials* **2011**, *32*, 1778–1786. [CrossRef]

21. Guillén, C.; Herrero, J. Polycrystalline growth and recrystallization processes in sputtered ITO thin films. *Thin Solid Films* **2006**, *510*, 260–264. [CrossRef]

22. Choi, S.K.; Lee, J.I. Effect of film density on electrical properties of indium tin oxide films deposited by dc magnetron reactive sputtering. *J. Vac. Sci. Technol. A Vacuum Surfaces Film.* **2001**, *19*, 2043–2047. [CrossRef]

23. Ryynänen, T.; Toivanen, M.; Salminen, T.; Ylä-Outinen, L.; Narkilahti, S.; Lekkala, J. Ion Beam Assisted E-Beam Deposited TiN Microelectrodes—Applied to Neuronal Cell Culture Medium Evaluation. *Front. Neurosci.* **2018**, *12*, 1–13. [CrossRef]
24. Nisch, W.; Böck, J.; Egert, U.; Hämmerle, H.; Mohr, A. A thin film microelectrode array for monitoring extracellular neuronal activity in vitro. *Biosens. Bioelectron.* **1994**, *9*, 737–741. [CrossRef]
25. Kim, R.; Nam, Y. Electrochemical layer-by-layer approach to fabricate mechanically stable platinum black microelectrodes using a mussel-inspired polydopamine adhesive. *J. Neural Eng.* **2015**, *12*, 026010. [CrossRef] [PubMed]
26. Hyvärinen, T.; Hyysalo, A.; Kapucu, F.E.; Aarnos, L.; Vinogradov, A.; Eglen, S.J.; Ylä-Outinen, L.; Narkilahti, S. Functional characterization of human pluripotent stem cell-derived cortical networks differentiated on laminin-521 substrate: Comparison to rat cortical cultures. *Sci. Rep.* **2019**, *9*, 1–15. [CrossRef] [PubMed]
27. Rossnagel, S.M.; Mikalsen, D.; Kinoshita, H.; Cuomo, J.J. Collimated magnetron sputter deposition. *J. Vac. Sci. Technol. A Vacuum Surfaces Film.* **1991**, *9*, 261–265. [CrossRef]
28. George, S.M. Atomic layer deposition: An overview. *Chem. Rev.* **2010**, *110*, 111–131. [CrossRef]
29. Blau, A. Cell adhesion promotion strategies for signal transduction enhancement in microelectrode array in vitro electrophysiology: An introductory overview and critical discussion. *Curr. Opin. Colloid Interface Sci.* **2013**, *18*, 481–492. [CrossRef]
30. Meng, L.-J.; Gao, J.; Silva, R.A.; Song, S. Effect of the oxygen flow on the properties of ITO thin films deposited by ion beam assisted deposition (IBAD). *Thin Solid Films* **2008**, *516*, 5454–5459. [CrossRef]

micromachines

Article

Buoyancy-Free Janus Microcylinders as Mobile Microelectrode Arrays for Continuous Microfluidic Biomolecule Collection within a Wide Frequency Range: A Numerical Simulation Study

Weiyu Liu [1], Yukun Ren [2],*, Ye Tao [2], Hui Yan [3],*, Congda Xiao [1] and Qisheng Wu [1]

[1] School of Electronics and Control Engineering, Chang'an University, Middle-Section of Nan'er Huan Road, Xi'an 710064, China; liuweiyu@chd.edu.cn (W.L.); 2016904117@chd.edu.cn (C.X.); qshwu@chd.edu.cn (Q.W.)

[2] State Key Laboratory of Robotics and System, Harbin Institute of Technology, West Da-zhi Street 92, Harbin 150001, China; tarahit@gmail.com

[3] School of Mechatronics Engineering, Harbin Institute of Technology, West Da-zhi Street 92, Harbin 150001, China

* Correspondence: rykhit@hit.edu.cn (R.Y.); yanhui@hit.edu.cn (H.Y.); Tel.: +86-0451-8641-8028 (Y.R.)

Received: 3 February 2020; Accepted: 9 March 2020; Published: 10 March 2020

Abstract: We numerically study herein the AC electrokinetic motion of Janus mobile microelectrode (ME) arrays in electrolyte solution in a wide field frequency, which holds great potential for biomedical applications. A fully coupled physical model, which incorporates the fluid-structure interaction under the synergy of induced-charge electroosmotic (ICEO) slipping and interfacial Maxwell stress, is developed for this purpose. A freely suspended Janus cylinder free from buoyancy, whose main body is made of polystyrene, while half of the particle surface is coated with a thin conducting film of negligible thickness, will react actively on application of an AC signal. In the low-frequency limit, induced-charge electrophoretic (ICEP) translation occurs due to symmetric breaking in ICEO slipping, which renders the insulating end to move ahead. At higher field frequencies, a brand-new electrokinetic transport phenomenon called "ego-dielectrophoresis (e-DEP)" arises due to the action of the localized uneven field on the inhomogeneous particle dipole moment. In stark contrast with the low-frequency ICEP translation, the high-frequency e-DEP force tends to drive the asymmetric dipole moment to move in the direction of the conducting end. The bidirectional transport feature of Janus microspheres in a wide AC frequency range can be vividly interpreted as an array of ME for continuous loading of secondary bioparticles from the surrounding liquid medium along its direction-controllable path by long-range electroconvection. These results pave the way for achieving flexible and high-throughput on-chip extraction of nanoscale biological contents for subsequent on-site bioassay based upon AC electrokinetics of Janus ME arrays.

Keywords: induced-charge electrokinetic phenomenon; ego-dielectrophoresis; mobile electrode; Janus microsphere; continuous biomolecule collection; electroconvection

1. Introduction

Lab-on-a-chip technology requires the development of new methods to manipulate small fluid and particle entities at micrometer dimension [1,2]. Discrete electrode array embedded in microfabricated fluidic networks stands for a brand new hope for direct electrokinetic actuation either on liquid suspension [3–6] or solid particles [7–9] dispersed in the fluid. Electrokinetics (EK) and electrohydrodynamics (EHD) of leaky dielectric medium [10,11] in microsystems have received unprecedentedly increasing attention from the microfluidic society for the last two decades. Traditional linear electroosmosis (EO) [12], electrowetting on dielectrics [13], induction EHD [14], injection EHD [15], conduction EHD [16], as well as nonlinear electroosmosis [17] are all typical representatives of physical phenomena where an external electric field is applied for driving fluids on nanoliter scale.

The common trait of EK and EHD is characterized by an active interaction between local electric field and the space charge cloud induced by itself to exert net electrostatic body forces that drive the motion of liquid medium, suspending colloids, discrete droplets as well as biological content of the microfluidic system, in the context of the so-called Ohm model [18–20] in close connection with electrochemical polarization [21–24] at a charged solid/electrolyte interface. The fast advance of microfabrication technique during the last ten years has allowed for an ease with which conducting metal plates can be patterned and inserted into microfluidic devices. For such, DC and AC electric fields as well as their delicate combinations have been widely employed for manipulating particle and liquid contents of microsystems [25,26].

In terms of the physical origin of the relevant charge layers, the primary electrical force exerting on microscale solid entities suspended in aqueous electrolytes includes electrophoresis (EP) [27–29], nonlinear induced-charge electrophoresis (ICEP) [30–35], and dielectrophoresis [36–38]. Directed "force-free" electrophoretic delivery of microspheres in a constant DC electric field is caused by the linear electroosmosis fluid stress on the surface of the natively charged particles, which is due to the interaction between an imposed tangential field and the natural diffuse screening cloud adjacent to the target sample suspended in aqueous electrolyte [39].

Unlike EP, both ICEP and dielectrophoresis (DEP) depend quadratically on the applied voltage, so they are nonlinear electrokinetic effects, wherein an applied electric field acts on its own induced charge to engender even time-averaged electrostatic particle motion under AC forcing. In ICEP, the induce charge cloud inside the electric double layer (EDL) is due to electrochemical polarization of ideally polarizable surfaces under the influence of an external electric field [34]. Different from ICEP, DEP arises from the bipolar surface charge, both free and bound, induced at the particle/solution interface via the mechanism of Maxwell-Wagner structural polarization [40,41].

In general, for symmetric particle entities, DEP and ICEP can only occur in a field gradient, so that symmetry breaking in electrical polarization induces a net electrokinetic motion along or against the gradient of electric field strength [10]. In fact, DEP and ICEP always tend to counterbalance one another, and gold spheres should keep stationary in DC limits from a theoretical perspective [42]. Nevertheless, dipolophoresis takes place in static fields in practical experiments; in that the present theory of ICEP would routinely overestimate the ICEO flow velocity by about one to two orders of magnitude, resulting in the dominating role of DEP over ICEP [43,44].

The intricate interplay between DEP and ICEP in dipolophoresis of symmetric metal colloids in field gradients makes dipolophoresis of ideally polarizable particles quite impractical for application in real experiments. So, great attention has been paid to the ICEP translation of Janus colloids during the last decade [33,45–48]. A Janus particle, whose half body is far less polarizable and another half is far more polarizable than the liquid suspension, serves as a typical example of solid entity with natural symmetry breaking in geometry and/or electrical properties [49]. In this way, ICEP translation of Janus colloids readily occurs in a uniform electric field even in the absence of a field gradient [50], which renders them good candidates for serving as mobile electrodes. Of course, such conducting electrodes embedded in microfluidic channels can also be fabricated using alternative materials such as liquid metal, as has been well studied by pioneering researchers [51–56].

It has been reported that, for a Janus microsphere subjected to a uniform low-frequency AC forcing, symmetry breaking in ICEO would first induce a net rotating motion due to combined ICEP stress and electro-orientational (ER) torque (See Supporting Video S1). A steady state is reached when the hemisphere interface aligns with the field axis, with the specific azimuth angle determined by its initial orientation. Then, the strong ICEO vortex flow field around the conducting end would push the Janus particle to transport unidirectionally with its insulating end moving ahead in the low-frequency limit, as shown in Figure 1, which is perpendicular to the AC forcing.

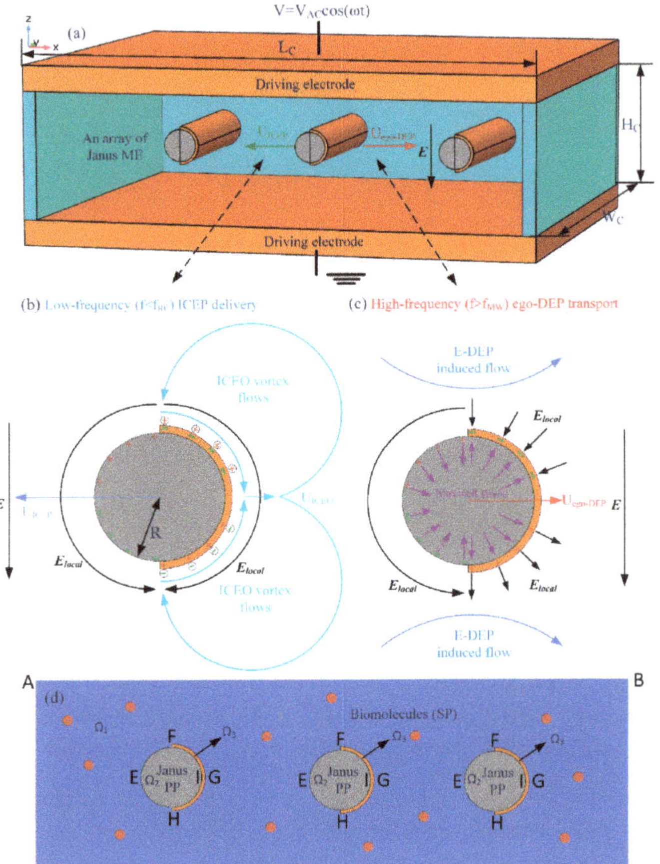

Figure 1. A schematic diagram of the direction-controllable electrokinetic transport of an array of cylindrical Janus mobile microelectrodes (ME) freely suspended in aqueous electrolyte under a uniform AC electric field, with the specific geometric size given in Table 1. (**a**) A 3D schematic diagram of the buoyancy-free Janus ME array that can move unidirectionally in perpendicular orientation to the externally-imposed AC forcing within a wide frequency range. (**b**) At frequencies below the inverse RC time scale for electrochemical polarization on the ideally polarizable surface of thin gold membrane, the induced-charge electrophoretic (ICEP) effect associated with induced-charge electroosmotic (ICEO) vortex flow field on the conducing end propels electrohydrodynamically the Janus ME to move in the direction of the insulating end. (**c**) At frequencies approaching and even exceeding the reciprocal charge relaxation time of the bulk electrolyte, ICEO vanishes due to electrochemical ion relaxation, and a net Maxwell stress from Maxwell-Wagner interfacial polarization renders the Janus ME array move electrokinetically in the direction of the conducting end, namely, the phenomenon of "ego-dielectrophoresis (e-DEP)". (**d**) An illustrative simulation domain for the electrokinetic motion of buoyancy-free Janus microspheres (primary particle, PP) as mobile microelectrode arrays for microfluidic biomolecule (secondary particle, SP) collection within a wide frequency range. It is noteworthy that the background AC electric field is uniform and there is no field gradient far from the Janus ME array.

The research group of Yossifon has recently discovered that a Janus colloid can also move perpendicular to the applied field for frequency even higher than the inverse RC time constant for electrochemical polarization of the conducting side, a phenomenon termed "self-dielectrophoresis (s-DEP)" [57]. More importantly, the potential use of such active particles as carriers of both organic and inorganic biological cargo was recently discovered and reported by the same great group as well [58,59]. In s-DEP, an asymmetric microsphere always sediments to the channel floor due to a large mass density of the conducting hemisphere in comparison to that of the liquid suspension. So, a strong electric field intensity is produced inside the narrow gap between the bottom section of the particle and the channel floor. This localized field gradient interacts actively with the asymmetric induced dipole moment, and thereby produces a net time-averaged electrostatic force that tends to transport the Janus colloid with its conducting side moving ahead [58], resulting in a reversal in the translating motion due to s-DEP with respect to that driven by ICEP. For s-DEP to occur, nevertheless, besides sinking to the channel bottom surface, the field frequency is strictly confined to be no more than the reciprocal Debye relaxation time of the liquid bulk.

To address the above issue, we reinvestigated the frequency-dependence of the translating behavior of Janus colloids in a unique configuration of microfluidic device, wherein a pair of face-to-face oppositely polarized parallel-plate electrode sandwiches a central microscale fluidic channel of a rectangular cross section (Figure 1a). To keep the cost of the memory within our available computer resource, a 2D calculation domain representing the x-z cross section was considered for the numerical simulation, wherein a Janus cylinder or an array of Janus mobile microelectrodes (ME) is suspended in a conducting buffer medium with its long axis coinciding with the y axis. Different from previous cases, the solid Janus entity in current study has a uniform body made of polystyrene (PS), and the surface of its half body is coated with a thin film of ideally polarizable noble metal material (such as gold). So, in the present case, it can be assumed that the Janus ME is freely suspended in water electrolyte, and thereby no sedimentation occurs.

In essence, the phenomenon of s-DEP would not happen without a close contact of the asymmetric particle dipole moment with the insulating substrate. In the current situation, however, the Janus entity is neutral in buoyancy effect, and thereby it is not possible for the physical condition of s-DEP-induced translation to be satisfied. In fact, up to now, to the best of our knowledge, all the literatures regarding electrokinetics of Janus particles (JP) focused on the motion of JPs near the wall or finally move close to the channel bottom surface. That is, there is still no such experimental data about the electrokinetic behavior of JP far away from the wall. In order to confirm whether or not a directed motion akin to s-DEP can still occur for a freely suspended Janus entity, its electrokinetic behavior as a function of time is numerically calculated by using a fully coupled mathematical model with a transient solver. In the current model, both the effects that induced-charge electroosmotic (ICEO) flow and Maxwell-Wager interfacial polarization have on the fluid-structure interaction are taken into account, and we run the simulation in a broad frequency range from DC limit to beyond the charge relaxation time of the liquid bulk. A brand new electrokinetic translating behavior, called "ego-DEP (e-DEP)" is found in the high-frequency limit. Unlike s-DEP, e-DEP requires neither a close contact between the particle and substrate surface, nor a limit of high-frequency range to be no more than the inverse Debye relaxation time. The discovery of the phenomenon of e-DEP enriches the electrodynamic behavior of buoyancy-free asymmetric Janus mobile microelectrodes (ME), which holds great potential in on-chip biomedical applications.

2. Methods

2.1. Device Configuration

The geometry of the microfluidic device is shown Figure 1a. The cubic fluidic chamber has a length of L_C, a width of W_C, and a height of H_C, with the specific channel dimensions shown in Table 1. The top and bottom walls of the channel consist of a pair of parallel conducting plates (indicated by the orange color), while all other sidewalls are frequency-independent ideal insulators in comparison with the leaky dielectric saline solution.

Table 1. Geometric size of the fluidic device shown in Figure 1.

Symbol	Implication	Value
H_C	Channel height	50 μm
L_C	Channel length	300–600 μm
W_C	Channel Width	500 μm
R	Radius of the Janus ME	3–9 μm
H_{mem}	Gold membrane thickness	50 nm
L_G	Interparticle separation	50 μm
r	Radius of the biomolecules	50 nm

A microscale Janus cylinder or an array of such Janus mobile microelectrodes (ME), with the long axis orienting perfectly along the y dimension, is freely suspended in the central level of the thin liquid layer. Half of its PS body is in direct contact with the electrolyte, while another half has a thin coating layer of ideally polarizable gold material of a small but finite thickness of H_{mem} = 50 nm that inhibits sedimentation. The polar interface bisecting the asymmetric cylindrical entity has a specific azimuth angle θ with respect to the horizontal x-axis. This kind of Janus cylinder can be chemically synthesized by bulk phase process on an industrial scale as compared to those methods based on the use of interfaces to break the symmetry [60]. It is found that the half PS body with thin coatings of gold membrane is an almost equal-potential half body under external AC forcing due to an induction process akin to electrostatic screening (not shown). So, e-DEP is able to still occur for a complete more polarizable hemisphere. However, e-DEP requires the JP to be freely suspended in the electrolyte solution without sedimentation, so that we prefer to make use of the thin coating case throughout this work.

On application of a standing AC voltage signal $V = V_{AC}\cdot\cos(\omega t)$, a possibly inclined Janus cylinder will react by combined ICEP and e-DEP. It will first realign its central polar interface in the direction of the electric field by a torque of an electric origin, then transport unidirectionally with its insulating or conducting end moving ahead, as depending on whether the applied field frequency is in the low (ICEP) or high (e-DEP) frequency range. The bi-directional electrokinetic behavior can be exploited for continuous loading and collection of secondary biomolecules as the array of Janus ME transports unidirectionally within the channel (Figure 1d).

2.2. Mathematical Model

2.2.1. AC Electric Field

The whole computational domain can be divided into four interrelated subregions, including the electrolyte bulk, the PS body of the Janus colloid, the thin gold-film coating, as well as the induced double layer (IDL) formed at the interface between the conducting film and liquid medium. The former three are treated as bulk zones, while the latter is represented by a conjugating condition of a sharp voltage drop across the IDL according to charge conservation equation. In the quasi-electrostatic limit, the electric potential field $\widetilde{\phi}$ in all the three bulk regions is governed by the current continuity condition in the sinusoidal steady state [61–63]:

$$\nabla \cdot \left((\sigma_i + j\omega\varepsilon_i)\nabla\widetilde{\phi}_i\right) = 0 \text{ for } i = 1, 2, 3 \tag{1}$$

where the subscripts i = 1, 2, 3 represent the electrolyte, PS material, and the gold membrane in sequence (Figure 1d). σ and ε denote the electric conductivity and dielectric permittivity of the corresponding material. With uniform electrical properties for each domain, the control equations of the electric fields are reduced to the Laplace equations with zero induced charge [64–66]:

$$\nabla^2\widetilde{\phi}_i = 0 \text{ for } i = 1, 2, 3 \tag{2}$$

At the interface between the liquid medium and the bare PS body, we have the following conjugating conditions [67]:

$$(\sigma_1 + j\omega\varepsilon_1)\nabla\widetilde{\phi}_1 \cdot \boldsymbol{n} = j\omega\varepsilon_2\nabla\widetilde{\phi}_2 \cdot \boldsymbol{n}$$
$$\widetilde{\phi}_1 = \widetilde{\phi}_2$$

(3)

where we have assumed zero bulk conductivity of the PS main body. The polarizability of the PS material is too low to induce observable electrochemical polarization at the PS/medium interface. At the PS/gold interface, similar conjugating conditions are applied:

$$j\omega\varepsilon_2\nabla\widetilde{\phi}_2 \cdot \boldsymbol{n} = \sigma_3\nabla\widetilde{\phi}_3 \cdot \boldsymbol{n}$$
$$\widetilde{\phi}_2 = \widetilde{\phi}_3$$

(4)

At the membrane/electrolyte interface, the voltage drop across the induced double layer is related to the normal electric field right outside the Debye layer according to electric current continuity [68–74]:

$$(\sigma_1 + j\omega\varepsilon_1)\nabla\widetilde{\phi}_1 \cdot \boldsymbol{n} = j\omega C_0\left(\widetilde{\phi}_1 - \widetilde{\phi}_3\right)$$
$$\sigma_3\nabla\widetilde{\phi}_3 \cdot \boldsymbol{n} = j\omega C_0\left(\widetilde{\phi}_3 - \widetilde{\phi}_1\right)$$

(5)

where we have invoked the classical linear RC-circuit theory of field-induced Debye screening, under the limit of small double-layer voltage drop and thin boundary layer. $C_0 = C_D/(1 + \delta)$ is the total IDL capacitance, which is in effect a series connection of the diffuse layer capacity $C_D = \varepsilon_1/\lambda_D$ and the Stern layer capacity C_S, in terms of the Debye screening length $\lambda_D = (D\varepsilon_1/\sigma_1)^{1/2}$ and the surface capacitance ratio of $\delta = C_D/C_S$.

Only a part of the total double-layer voltage drops across the diffuse layer of mobile ions, namely, the induced zeta potential $\widetilde{\zeta} = \widetilde{\phi}_3 - \widetilde{\phi}_1/1 + \delta$. This contributes to a sharp velocity discontinuity at the membrane surface due to the action of induced-charge electroosmotic flow [75–77]:

$$\left\langle u_{\text{slip}} \right\rangle = -\frac{\varepsilon_1}{2\eta}\text{Re}\left(\widetilde{\zeta}\widetilde{E}_t^*\right) = \frac{\varepsilon_1}{2\eta(1+\delta)}\text{Re}\left(\left(\widetilde{\phi}_1 - \widetilde{\phi}_3\right)\left(\widetilde{E}_1 - \widetilde{E}_1 \cdot \boldsymbol{n} \cdot \boldsymbol{n}\right)^*\right)$$

(6)

where $\eta = 0.001$ Pa·s denotes the dynamic viscosity of the aqueous suspension.

On the conducting surface of a pair of face-to-face electrode plates, double-layer polarization is omitted considering the specific working frequency band, so fixed AC voltage phasor is imposed to establish a time-harmonic voltage difference along the channel height direction (Figure 1a):

$$\widetilde{\phi}_1 = V_1 \text{ at the top electrode plate } \widetilde{\phi}_1 = 0 \text{ at the bottom counterpart}$$

(7)

At the two end boundaries of the computational domain, the normal component of the total electric current vanishes to mimic a finite calculation volume:

$$\nabla\widetilde{\phi}_1 \cdot \boldsymbol{n} = 0$$

(8)

2.2.2. Particle Motion

The Janus entity is treated as linear elastic material; its displacement S is governed by its solid stress tensor $\overleftrightarrow{T}_S(S)$:

$$\nabla \cdot \overleftrightarrow{T}_S(S) + \boldsymbol{f} = 0$$

(9)

where f is a source term of body force density acting on the asymmetric particle. In current analysis, we exclude the existence of any kind of body force by imposing $f = 0$, since the particle is freely suspended in the electrolyte solution with an almost neutral mass density. The Cauchy stress of the solid phase $\overleftrightarrow{T}_S(S)$ is clearly a function of the displacement S of the Janus ME.

Considering that the Janus entities are incompressible neo-Hookean material, which can be described by the strain energy density function, as given by:

$$W = G_0(I_C - 3)/2 \tag{10}$$

where G_0 is the shear modulus of the intrinsically deformable particles, $I_C = \text{tr}(\mathbf{C})$ denotes the first invariant of the right Cauchy–Green tensor. $\mathbf{C}(E, v) = \mathbf{F}_T\mathbf{F}$, with E the Young modulus, v the Poisson's ratio, $\mathbf{F}$ the deformation gradient tensor that can be computed by $\mathbf{F} = \nabla\mathbf{S} + \mathbf{I}$.

The corresponding Cauchy stress of the neo-Hookean material can be described by:

$$\overset{\leftrightarrow}{T}_S(S) = J^{-1}\mathbf{PF}^T \tag{11}$$

where J is the determinant of the deformation gradient tensor $\mathbf{F}$, $J = 1$ for an incompressible Neo-Hookean material, and $\mathbf{P} = \partial Ws/\partial\nabla_X S$ is the first Piola–Kirchhoff stress.

Since the Janus ME used in current analysis is rigid and non-deformable, we arbitrarily set the Young modulus as large as 10^{12} Pa, while the Poisson's ratio has a normal value of 0.33 to approach a rigid body. In this way, our model can be easily extended to describe the electrokinetic motion of deformable Janus cylinders with different shear modulus. That is, we make use of the model for deformable particles to equivalently describe the fluid-structure interactions of rigid Janus ME arrays by setting an extremely high Young modulus, which is feasible according to the correct simulation results in the section of numerical validation (Section 3.1).

2.2.3. Fluid Motion

We solve the Navier–Stokes equation for incompressible Newtonian fluid to acquire the transient flow field of the liquid suspension:

$$\rho\frac{\partial u_f}{\partial t} + \rho\big(u_f \cdot \nabla\big)u_f = \nabla \cdot \left(-p\overset{\leftrightarrow}{I} + \eta\Big(\nabla u_f + \nabla u_f^T\Big)\right) \quad \nabla \cdot u_f = 0 \tag{12}$$

where ρ denotes the liquid mass density, p the hydraulic pressure, and u_f the flow velocity field.

2.2.4. Electric-Field-Mediated Particle-Fluid Interaction

At the particle/fluid interface, the physical constraint of total stress balance implies:

$$\overset{\leftrightarrow}{T}_S(S) \cdot n + \overset{\leftrightarrow}{T}_H \cdot n + \left\langle\overset{\leftrightarrow}{T_E}\right\rangle \cdot n = 0 \tag{13}$$

where $\overset{\leftrightarrow}{T}_S(S)$ is the solid stress tensor, $\overset{\leftrightarrow}{T}_H = -p\overset{\leftrightarrow}{I} + \eta\Big(\nabla u_f + \nabla u_f^T\Big)$ the hydrodynamic stress tensor, and $\left\langle\overset{\leftrightarrow}{T_E}\right\rangle = \frac{1}{2}\varepsilon_1\text{Re}\big(EE^* - \frac{1}{2}E \cdot E^*I\big)$ the time-averaged Maxwell stress tensor evaluated along the particle surface on the liquid side.

A slip velocity appears on the surface of conducting membrane, while a continuity of the system speed is maintained at the liquid/PS interface:

$$u_f = u_s + \frac{\varepsilon_1}{2\eta(1+\delta)}\text{Re}\Big(\big(\widetilde{\phi}_1 - \widetilde{\phi}_3\big)\big(\widetilde{E}_1 - \widetilde{E}_1 \cdot n \cdot n\big)^*\Big)$$
$$u_f = u_s \tag{14}$$

where $u_s = \partial S/\partial t$ is the local velocity of the Janus ME.

2.3. Numerical Simulation

A commercial software package, Comsol Multiphysics (version 5.3a, COMSOL Inc., Stockholm, Sweden), is used herein for numerically calculating the transient motion behavior of the asymmetric

Janus cylinder immersed in aqueous electrolyte driven by an external AC forcing. On account of its time-dependent nature, strong fluid-structure interaction has to be incorporated into the simulation model. The 2D computational domain (Figure 1d) is virtually a x-z cross section of the 3D schematic diagram displayed in Figure 1a, which consists of an array of circular Janus ME freely suspended in a straight liquid channel sandwiched between a pair of AC-powered parallel electrode plates. The specific simulation procedure is as below.

(1) At first, we compute the Laplace equation (Equation (2)) for the three separate regions, including the bulk electrolyte, PS material, and the thin gold film, which is subjected to those conjugating conditions (Equations (3)–(5)) at different sharp material interfaces. In the meantime, AC voltage phasor is prescribed on the oppositely polarized metal plates (Equation (7)), and the normal potential gradient vanishes on both ends of the calculation domain (Equation (8)).

(2) Once the AC potential is known, it is then possible for us to get the knowledge of the particle–fluid interaction by computing Equation (9) in the solid region and Equation (12) in the liquid region, respectively. Stress balance incorporating the electrostatic force is imposed at the particle/liquid interface (Equation (13)). The speed of movement is continuous on the insulating surface of the Janus cylinder (Equation (14)), while a sharp ICEO slipping appears on the gold coating layer (Equation (14)). In this way, the combined effect of ICEO slipping and DEP force on the electrokinetic behavior of the Janus ME array is naturally included in our simulation model.

(3) Once the solid motion and flow field due to the action of AC electrokinetics are known, a particular moving mesh scheme is used to trace the transient evolution of the solid/liquid contact interface, with a request of the mesh velocity being equal to the computed particle velocity in the normal direction of the surface of the Janus ME. Once the structural interface moves more or less, all the field variables have to be recalculated with the simulation results regenerated incessantly by repeating (1)–(2) in the next round of numerical computation. Consequently, the calculation time step should be orders of magnitude shorter than the characteristic time scale of the solid and fluid mechanics, which is cautiously defined in the settings of the solver.

Since the AC electric field is complex in essence, while other field variables are real numbers, the AC electrostatic potential, and the field-induced fluid-solid interaction are solved with a transient solver in a segregating coupled way. The size of mesh on the particle surface is required to be no more than one-twentieth of the diameter of the rigid Janus cylinder, which remains a circular shape in the 2D domain all the time. This meshing scheme leads to 30,000–90,000 triangular grids in the 2D space, as determined specifically by the size and position of the Janus ME of interest. A stop condition is imposed to the transient solver, and dictates that the simulation process comes to a controlled halt once the leftmost (or the rightmost) end of the Janus ME approaches the entrance (or the exit) of the device channel of a rectangular shape. Please refer to Table 2 for the boundary conditions, and Table 3 for the simulation parameters pertinent to the fluidic device taking a Janus ME array as the functional microelectrodes.

Table 2. Boundary conditions for the simulation domain shown in Figure 1d implemented in Comsol Multiphysics.

Boundary Index	Electric Field Equation (2) $(\Omega_1 + \Omega_2 + \Omega_3)$	Flow Field Equation (12) (Ω_1)	Solid Mechanics Equation (9) $(\Omega_2 + \Omega_3)$	Concentration Field Equation (15) (Ω_1)
A-C	Zero current flux	Zero normal stress	N/A	Open boundary
A-B	Equation (7)	No slip	N/A	No flux
B-D	Zero current flux	Zero normal stress	N/A	Open boundary
C-D	Equation (7)	No slip	N/A	No flux
F-E-H	Equation (3)	Equation (14)	Equation (13)	No flux
F-I-H	Equation (4)	N/A	N/A	N/A
F-G-H	Equation (5)	Equation (14)	Equation (13)	No flux

Table 3. Simulation parameters pertinent to the microfluidic device taking advantage of an array of Janus mobile microelectrodes for continuous loading and collection of surrounding biomolecules for potential biomedical applications.

Symbol	Implication	Value
σ_f	Liquid conductivity	0.001 (ICEP)–0.1 S/m (e-DEP)
σ_{gold}	Membrane conductivity	10^7 S/m
σ_{ps}	Conductivity of polystyrene	0 S/m
ε_f	Liquid permittivity	$80\varepsilon_0$
ε_{gold}	Gold permittivity	$10\varepsilon_0$
ε_{ps}	Permittivity of polystyrene	$3\varepsilon_0$
ε_0	Vacuum permittivity	8.85×10^{-12} F/m
Dion	Diffusivity of ions	2×10^{-9} m^2/s
f_{RC}	RC charging frequency	1.7–20.9 kHz
f_{MW}	Debye relaxation frequency	225 kHz–22.5 MHz
τ_{RC}	RC time constant	7.62×10^{-6}–9.19×10^{-5} s
τ_{MW}	Debye time constant	7.1×10^{-9}–7.1×10^{-7} s
λ_D	Debye screening length	3.76–37.6 nm
C_S	Stern layer capacity	0.8 F/m^2
C_D	Diffuse layer capacity	0.0188–0.188 F/m^2
C_0	Initial molecule concentration	1 nM
T	Ambient temperature	293.15 K
D	Diffusivity of nanoparticles	4.3×10^{-12} m^2/s
ρ	Liquid mass density	1000 kg/m^3
η	Liquid dynamic viscosity	0.001 Pa·s
V_1	Voltage amplitude	1–2 V
f	Signal frequency	100 Hz–500 MHz
E	Young modulus	10^{12} Pa
ν	Poisson's ratio	0.33

3. Results and Discussion

3.1. Model Validation

Before making any quantitative calculations, we should demonstrate the validness of current physical model for dipolophoretic motion of asymmetric colloids at a given voltage. On the one hand, in the low-frequency limit, the Janus cylinder with an initial azimuth angle $\theta_0 = 50°$ relative to the horizontal direction first rotates counterclockwise by ICEP torque, which makes its polar interface dividing the two hemispheres of different polarizability align well with the imposed field lines. Then, it moves almost unidirectionally upstream in the direction of its insulating PS end. Lateral motion occurs at the same time, however, since ICEO vortices on the particle surface near a conducting wall will always attract it to approach the ideally polarizable surface (See Supporting Video S1).

On the other hand, at frequencies beyond the inverse Debye relaxation time of the liquid bulk, the inclined Janus ME will still rotate counterclockwise by ER torque and achieve an identical alignment, while with its conductive end moving ahead toward downstream (see Supporting Video S2). Under this situation, however, no evident deviation of the particle from the channel centerline can be observed, since the phenomenon of particle-wall DEP interaction is a near-field effect in stark contrast to the far-field ICEO streaming flows in the low-frequency condition. In this sense, the effectiveness of the simulation model developed herein is proved on a qualitative level. On this basis, we then make a detailed investigation on the AC electrokinetic behavior of an individual Janus cylinder or an array of Janus ME, by conducting direct numerical simulation incorporating the surface-coupled fluid–structure interaction with a transient solver.

3.2. Directed Electrokinetic Transport of Janus Entity in a Wide Frequency Range

3.2.1. AC Electrokinetics under the Low- and High-Frequency Limits

From now on, we focus on the directed transport behavior of a single Janus mobile electrode (ME), with the simplest configuration wherein its polar interface is in the direction of the external AC field lines at the early stage, as shown in Figure 2a–d. In the computation, suitable geometry dimension is selected for the electrokinetic chip: $L_C = 300$ μm, $H_C = 50$ μm, $R = 6$ μm, $\lambda_D = 37.6$ nm, along with the following physicochemical properties of the liquid and solid materials: $\sigma_f = 0.001$ S/m, $\sigma_{gold} = 107$ S/m, $\sigma_{ps} = 0$ S/m, $C_0 = 0.018$ F/m^2, $\varepsilon_f = 80\varepsilon_0$, $\varepsilon_{ps} = 3\varepsilon_0$, $\varepsilon_{gold} = 10\varepsilon_0$, $\tau_{RC} = C_D R/\sigma_f(1 + \delta) = 5.5 \times 10^{-5}$ s, $\tau_{MW} = \varepsilon_f/\sigma_f = 7 \times 10^{-7}$ s, and $\eta = 0.001$ Pa·s. The aqueous suspension is supposed to possess an electric conductivity of 0.001 S/m, giving rise to an inverse Debye relaxation time $f_{MW} = 1/2\pi\tau_{MW} = 225$ kHz of the bulk electrolyte, as well as a reciprocal RC time constant $f_{RC} = 1/2\pi\tau_{RC} = 2.9$ kHz for electrochemical polarization of the IDL at the membrane/suspension interface (Tables 1 and 3).

To begin with, we numerically studied two limiting cases of this subject, namely, the low-frequency ICEP delivery (Figure 2a,b), and the high-frequency e-DEP transport (Figure 2c,d) affected by a uniform AC electric field. In the parameter settings, the AC voltage amplitude is fixed at 1V. The electric field strength equals 20 V/mm, which is on the same order of magnitude with that applied in typical ICEO experiments.

As for the low-frequency situation, the field frequency is 100 Hz, being much less than the inverse RC time scale $f_{RC} = 2.9$ kHz for the field-induced Debye screening on the ideally polarizable surface of the conducting film. In this situation, the applied field lines bend across the particle/electrolyte interface around both hemispherical sides, in that the induced double layer (IDL) can be fully developed and then perfectly screen the normal field gradient emitted from the polarized gold membrane, as displayed in Figure 2b. As a consequence, the electric field pattern for the Janus cylinder at low frequencies is similar to that of a perfect insulating particle due to the occurrence of complete Debye screening on the conducting end. As shown in Figure 2a, since there is enough time for the bipolar counterions to amass within the thin boundary layer on the membrane surface, they are actively acted by the same frequency tangential AC forcing to engender a nonlinear ICEO vortex flow field around the conducting end, which even survives after time-averaging under AC (Figure 1b). ICEO slipping fluid motion injects from both the top and bottom ends, and then ejects selectively into the bulk suspension around the equatorial plane, resulting in the formation of a pair of ICEO micro-vortices in opposite rotating directions.

The induced-charge electro-osmotic flows, which originate from the action of an externally imposed AC electric field on its own induced charge within an induced double layer (IDL) on a polarizable solid surface immersed in electrolyte, won't converge. Instead, the flow streams are coming outward to form vortices (Figure 1b). This type of vortex flows is well documented in pioneering literatures, wherein the earliest researches on DC/AC-induced ICEO vortex flow field on the sharp corner singularity of dielectric blocks of a small but finite permittivity comparable to that of water suspension were reported [78–80]. These are indeed great works on time-averaged ICEO streaming vortices on dielectric solid surfaces (rather than ideal metal conductors).

Recently, we have also investigated a time-averaged ICEO vortex flow adjacent to the sharp-corner-singularity of leaky dielectric blocks of both a finite conductivity and permittivity in external time-harmonic AC forcing [81]. In this paper, under the thin layer approximation and small double-layer voltage drop, we deal with the IDL as a thin capacitive skin induced at the dielectric wall/electrolyte interface by the applied AC field. The interaction of the imposed tangential AC forcing with its own induced bipolar Debye screening charge within the IDL gives rise to a pair of ICEO vortex in counter-rotating directions around the corner of the dielectric wall. Besides, if an additional DC component is applied across the channel length direction, linear DC electroosmotic (DCEO) pumping and DC-ICEO vortex flow appear at the same time. Once the DC voltage is much larger than the AC counterpart, an evident unidirectional DCEO flow would superimpose on the ICEO vortex flow pair induced by the combined AC and DC forcing, which has a tendency to flatten out the upstream ICEO vortex while with the swirling shape of the downstream ICEO vortex keeping almost unvaried. To this end, according to the Newton third law, the ICEO vortex flow profile will propel the Janus cylinder to move in the direction of the insulating end, as indicated in Figure 2a wherein the nearly symmetric dipole moment is delivered by ICEP forcing toward upstream for a powering time of 0.1 s.

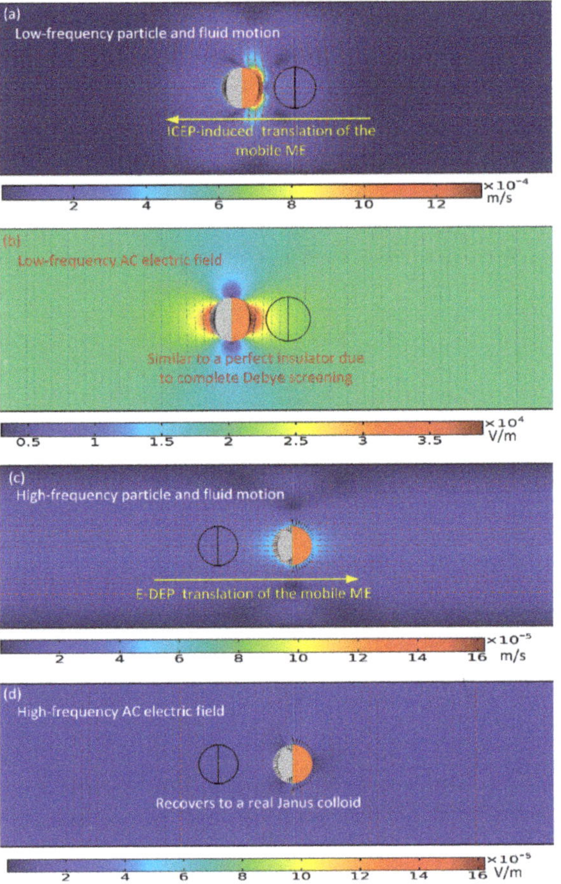

Figure 2. A simulation illustration of the electrokinetic delivery behavior of the Janus ME at distinct field frequencies under a given voltage amplitude of 1V. (**a**,**b**) In the low-frequency limit of f = 100 Hz and at a powering time of t = 0.1 s (See Supporting Video S1), (**a**) a surface and arrow plot of the ICEO-induced electrohydrodynamics (EHD) flow field and particle motion, and (**b**) an arrow and surface plot of the electric field around the Janus colloid of a symmetric induced dipole due to complete Debye screening on the conducting end. (**c**,**d**) In the high-frequency limit of f = 100 MHz and at a powering time of 1 s (See Supporting Video S2), (**c**) a surface and arrow plot of the ego-dielectrophoresis (e-DEP)-induced electrokinetic (EK) flow field and particle motion, and (**d**) an arrow and surface plot of the electric field around the Janus colloid of an asymmetric induced dipole due to the occurrence of electrochemical ion relaxation. The black arrows on the surface of the Janus particle (JP) in Figure 2 denote the interfacial Maxwell stress of a second-order voltage dependence.

For the sake of comparison, we then increase the imposed field frequency sharply from 100 Hz to 10 MHz. In this high-frequency limit, the IDL has already been short circuited by the constantly varying displacement current through it, so there is no longer asymmetric ICEO streaming flow around the solid entity. Since the permittivity effect dominates within the high frequency range, the insulating PS end still is less polarizable than the surrounding medium due to its lower dielectric permittivity compared to that of water. Nevertheless, the other hemisphere coated with thin gold membrane recovers to the intrinsic role of an ideal conductor on account of the incomplete capacitive charging of the IDL. In this way, the dipole moment of the conducting end is now positive, whereas the insulating end is still negative, which leads to the formation of an asymmetric induced dipole moment of the Janus ME (Figure 2d). Although the background electric field is uniform, the inhomogeneous dipole induced by the external field generates a strong asymmetry in electric field distribution around the particle. The action of the localized anisotropic field gradient due to the induced dipole acts on the asymmetric particle dipole moment itself, resulting in a new AC electrokinetic phenomenon called "ego-DEP (e-DEP)".

The schematic diagram of the interfacial induced charge and the net nonlinear Maxwell stress is qualitatively shown in Figure 1c, which helps us to understand the physics of e-DEP more easily to a great extent. From Figure 1c, the electrical force in the direction of the polar interface vanishes considering a symmetry in vertical polarization, while the net electrokinetic stress in perpendicular orientation to the polar interface has a net effect in trying to push the Janus ME to move horizontally in the direction of the conducting end. Besides, since both the induced surface charge (free plus bond) and localized field gradient change direction within each half AC cycle, the net e-DEP force survives well after time-averaging in harmonic forcing, and can thereby constantly transport the asymmetric dipole moment towards the conducting end. This can be evidenced by the simulation result in Figure 2c wherein the Janus ME moves downstream due to a net electrical stress that points in the direction of the conducting end.

To this end, it has been demonstrated that the Janus entity acting as active ME can be either delivered by ICEP rotating vortex at low frequency or transported by e-DEP in high frequency limit.

3.2.2. Electrokinetic Motion in a Wide Frequency Range

Although we have discovered a Janus ME can move electrokinetically in both low and high frequency limit. AC electrokinetic behavior in the intermediate frequency range is still unclear, however, so it is necessary for us to address this issue immediately. A parametric simulation study is then carried out to make it clear how the Janus cylinder makes a response to the applied AC forcing within a broad frequency range from 100 Hz to 100 MHz. Such an elaborate selection of the frequency limits is plausible, in that 100 Hz is much lower than the inverse RC time constant for capacitive charging of the IDL at the membrane/suspension interface, and 100 MHZ is much higher than the Debye relaxation frequency of the liquid bulk. This specific range defined by the lower and upper limits covers effectively the frequency band for any possible nonlinear electrokinetic phenomenon to occur.

As shown in Figure 3a, the direction of movement of the Janus ME is negative, and points to the insulating end from 100 Hz to 10 kHz, because ICEP propulsion dominates its motion behavior at frequencies around the inverse RC time scale f_{RC} = 2.9 kHz for electrochemical polarization of the membrane in direct contact with the aqueous electrolyte. The negative translational velocity equals -150 μm/s at 10 Hz, and gradually decreases to -30 μm/s with an increase of frequency to 10 kHz, which is due to an electrochemical ion relaxation of the double layer charge that weakens ICEO slipping velocity under higher excitation frequencies (Figure 3d).

On the other hand, for the imposed field frequency no less than 50 kHz, the moving direction of the Janus cylinder alters from negative to positive (Figure 3e). Under this circumstance, there is not enough time for the charged counterions to accumulate effectively within the thin boundary layer, so that the phenomenon of ICEO fades away, and the particle dipole moment recovers to an asymmetric one when considering there is no longer Debye screening on the conducting hemisphere. In this way, the e-DEP-enabled electrokinetic force governs the motion of the Janus cylinder, which increases with frequency and attains a plateau of 24 μm/s at frequencies beyond 1 MHz (Figure 3e). Any further increase of the field frequency exceeding 1MHz would no longer alter the magnitude of the induced dipole moment, in that the gold membrane has a conductivity of 10^7 S/m about ten orders of magnitude larger than that 10^{-3} S/m of the liquid suspension. Though the e-DEP-induced particle translating movement +24 μm/s (Figure 3e) within the high frequency range is smaller than −150 μm/s caused by ICEP in low frequency limit, it is still observable and provides a supplementary transport direction. The discovery of the effect of e-DEP enriches the AC electrokinetic behavior of Janus entity in a uniform electric field, and allows a direction-controllable delivery within a wide frequency range.

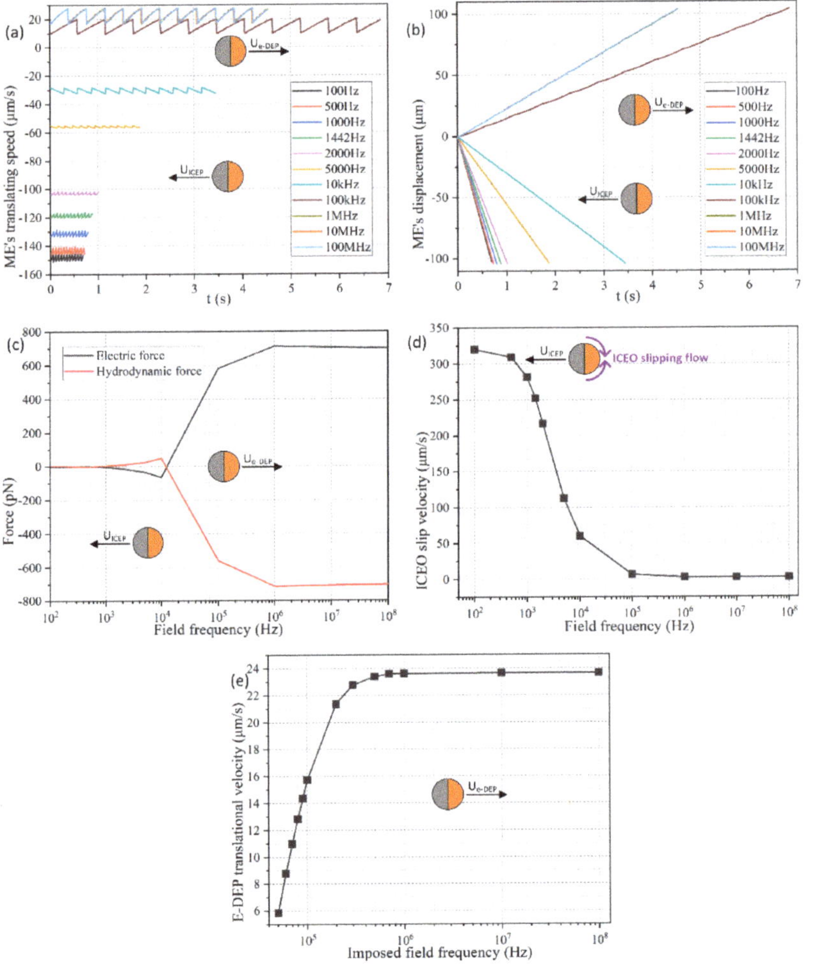

Figure 3. Influence of the imposed AC field frequency on the electrokinetic transportation behavior of an individual Janus ME with a diameter of 12 μm. (**a**) Time-dependent ME's translating speed under different signal frequencies for a same total displacement of 105 μm. (**b**) The horizontal displacement of the Janus colloid as a function of time in a wide frequency range. (**c**) The balance between the electric force and hydrodynamic force acting on the Janus entity as a function of the field frequency. (**d**) Frequency-dependent ICEO slipping velocity on the surface of the conducting end. (**e**) e-DEP-induced JP translating speed as a function of the imposed field frequency from 50 kHz to 100 MHz. It is noteworthy that ICEP dominates in the frequency range below 50 kHz, so the start threshold frequency is set at 50 kHz in Figure 3e.

Meanwhile, the transient moving speed behaves as an oscillating profile as time advances, which arises mainly from a dynamic balance among the various forces exerted on its surface, including active ICEP propelling, interfacial nonlinear Maxwell stress, Stokes drag force, as well as the stress from the rigid PS body, as can be clearly proved in Figure 3c wherein the electric force and hydrodynamic force always counterbalance one another. Even so, the horizontal displacement exhibits a monotonous growing trend as a function of time whatever the imposed field frequency is (Figure 3b). As a consequence, we pay less attention to the transient fluctuation in the speed of movement, and mainly focus on the time-averaged translational speed from now on.

3.3. Parametric Study for the Low-Frequency Induced-Charge Electrophoretic (ICEP) Translation

In this section, the field frequency is fixed at f = 100 Hz, which is well below the RC charging frequency of the IDL at the ideally polarizable surface of the gold membrane. In such a low-frequency limit, we discerned the effects that the dimension of the Janus entity and the magnitude of AC voltage have on the resulted particle translating kinematics.

As for the influence of the applied voltage, the colloidal radius is 6 μm, and the signal amplitude is arbitrarily swept from 1 V to 4 V with a fixed discrete interval of 1 V. As shown in Figure 4a, the ICEP motion in the direction of the insulating PS end has a translating speed that grows quadratically with the applied voltage, namely, $U_{ICEP} \propto V^2$ and thereby much less time is required for the Janus cylinder to cover an identical horizontal distance of 100 μm when the voltage is enhanced from 1 V to 4 V as other conditions remain unchanged. On the other hand, under a given voltage amplitude of 2 V, although the negative speed of movement also increases with increasing particle diameter, this growing trend has a power law with the exponent less than one as a function of colloidal dimension, that is, $U_{ICEP} \propto R^\alpha$, with $0 < \alpha < 1$. In this way, according to our direct simulation study, $U_{ICEP} \propto V^2 R^\alpha$ $(0 < \alpha < 1)$ is against the scaling rule of ICEO around a metal particle, $U_{ICEP} \propto V^2 R$, given by Bazant and Squires for an infinitely large liquid domain [82]. The reason behind this particular aberration is originated by a vertical confinement effect due to the finite space of current fluidic chamber, which merely has a height of 50 μm. Under this circumstance, as long as the particle diameter is no less than one-tenth of the channel vertical dimension (50 μm/10 = 5 μm), the vertical confinement effect will attempt to suppress the healthy development of the ICEO slipping vortex flow field adjacent to the conducing side (Figure 2a), which counteracts the enhanced induced polarization of the gold membrane from a larger particle radius, so that the scaling characteristic of the ICEP transporting speed eventually manifests as $U_{ICEP} \propto V^2 R^\alpha$ with $0 < \alpha < 1$.

The coupled interaction between the colloid dimension and AC voltage for the electrokinetic behavior of the Janus entity is displayed in Figure 4b,d as well, wherein we provide polynomial fitting for each data curve. The voltage-dependent moving velocity agrees well with the quadratic polynomial fitting (QPF) in Figure 4b, demonstrating the effectiveness of $U_{ICEP} \propto V^2$. Nevertheless, the radius-dependent translating speed is always lower than the linear polynomial fitting (LPF) in Figure 4d, which witnesses the correctness of the scaling law of $U_{ICEP} \propto R^\alpha$ $(0 < \alpha < 1)$.

To this end, within a fluidic chamber, in which as long as there is one dimension that is comparable to the size of the suspended particles, it is no longer convenient to increase the ICEP motion by enlarging the particle diameter due to the action of a confinement effect. Alternatively, enhancing the applied voltage serves as a more ideal way for accelerating the unidirectional delivery speed of the micro Janus cylinder in electrolytes.

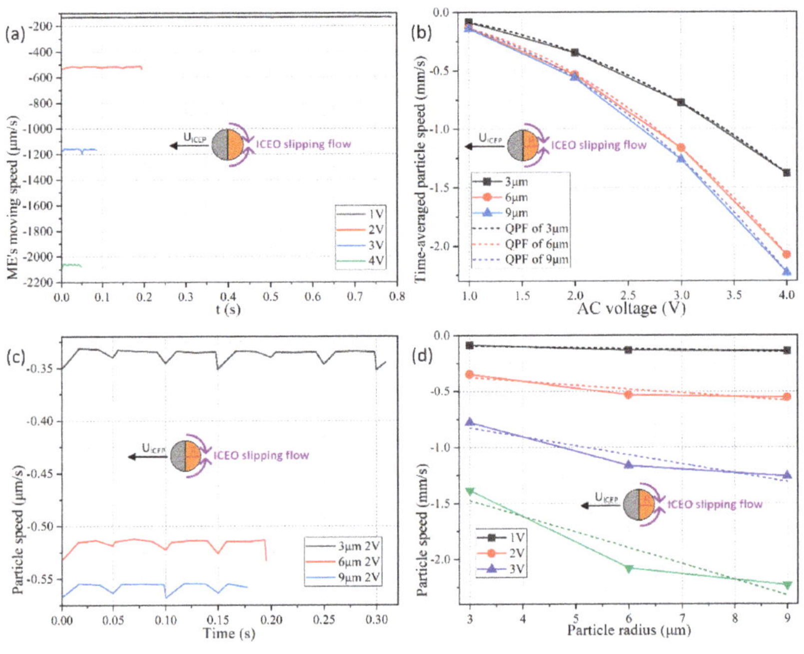

Figure 4. Parameter-dependence study of the low-frequency ICEP translation of the Janus ME in the direction of the insulating end at 100 Hz. (**a**) Moving speed of the Janus colloid as time elapses under different AC voltages at a given particle radius of 6 μm. (**b**) Voltage-dependence of the time-averaged particle's translating speed for different colloid radius as well as their quadratic polynomial fitting (QPF) curves. (**c**) Moving speed of the Janus colloid as time elapses for different particle radius at a given voltage of 2 V. (**d**) Diameter-dependence of the time-averaged particle's translating speed for different voltage amplitude as well as their linear polynomial fitting (LPF) curves.

3.4. Parametric Study For High-Frequency e-DEP Translation

To confirm the existence of e-DEP and its dependence on some key parameters, the field frequency is subsequently enhanced to 10 MHz. Under such high frequency excitations, both the electrochemical polarization at the membrane surface and the associated ICEP propulsion effect decay to zero due to electrochemical ion relaxation. So, only Maxwell-Wagner interfacial polarization plays an important role in high-frequency electrokinetic behavior of the Janus cylinder, with the schematic vividly shown in Figure 1c. The gold film recovers to its original role of a perfect conductor in the absence of conspicuous Debye screening. Consequently, symmetry breaking in the induced dipole moment of the Janus colloid takes place. Although a uniform background AC electric field is imposed along the channel depth direction (Figure 1a), the field strength is greatly perturbed to become quite uneven around the particle (Figure 2d), which acts on the asymmetric dipole induced by the field itself, so as to enable a continuous translating motion in the direction of the conducting end (Figure 2d).

As shown in Figure 5a, under a given colloid radius of $R = 6$ μm, as the voltage amplitude increases from 1 V to 4 V, the time-averaged moving speed of the Janus mobile electrode (ME) grows from 26.5 μm/s to 430 μm/s, implying a quadratic growth trend of the e-DEP-induced speed of movement as a function of the AC forcing. Namely, the scaling law of DEP motion in an inhomogeneous electric field above a microelectrode array, $U_{e\text{-}DEP} \propto V^2$, is still applied for e-DEP motion of the Janus ME affected by a uniform potential gradient, when considering its nonlinear nature that the localized field gradient interacts with its own induced asymmetric particle dipole moment.

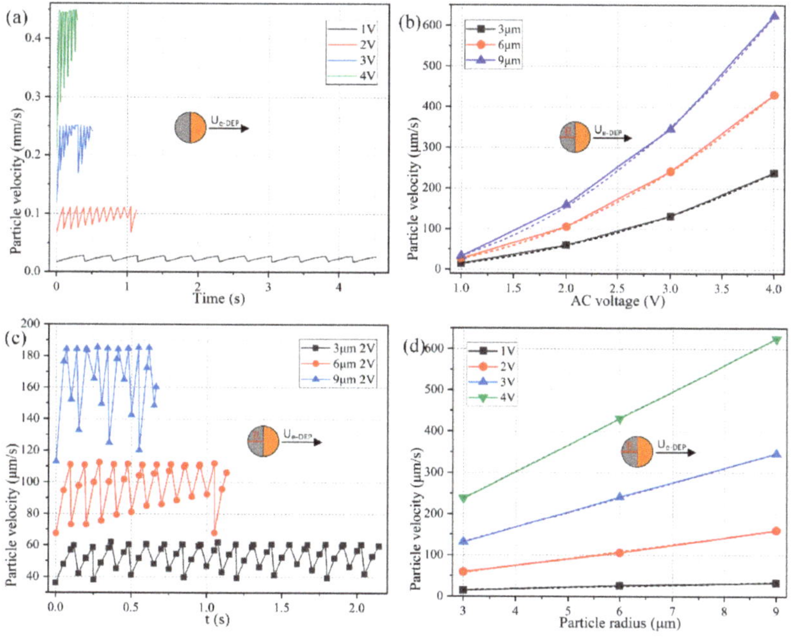

Figure 5. Parameter-dependence study of the high-frequency e-DEP translation of the Janus ME in the direction of the conducting end at 10 MHz. (**a**) e-DEP moving speed of the Janus colloid as time advances for different AC voltage with a given particle radius of 6 μm. (**b**) Voltage-dependence of the time-averaged particle's translating velocity for different colloid radius as well as their quadratic polynomial fitting (QPF) curves. (**c**) e-DEP moving speed of the Janus colloid as time advances for different particle radius at a given voltage of 2 V. (**d**) Diameter-dependence of the time-averaged particle's translating speed for different voltage amplitude as well as their linear polynomial fitting (LPF) curves.

On the other hand, the dependence of the e-DEP velocity on the dimension of the particle is somewhat counterintuitive. A shown in Figure 5c, the particle velocity at 10 MHz from e-DEP is almost linearly proportional to the radius of the ME, namely, $U_{e-DEP} \propto R$, which differs from the typical DEP velocity scale of $U_{DEP} \propto R^2$ in a field gradient. This disparity may be ascribed to the fact that, the convectional DEP motion is caused by a background field gradient, while the translating behavior of e-DEP is due to the action of a secondary field gradient around the particle actuated by a uniform AC forcing. As a consequence, the e-DEP speed obeys the scaling trait of $U_{e-DEP} \propto V^2 R$.

These conclusions can be also demonstrated in Figure 5b,d. As displayed in Figure 5b, the simulated data of the voltage-dependent e-DEP velocity (the solid lines) almost overlap with their QPF curves (the dashed lines). Unlike the vertical confinement effect that the finite channel height has on the ICEP translation in low-frequency limit (Figure 4d), this negative influence is not adaptable to the phenomenon of e-DEP at all. Even if the diameter of the ME increases to 18 μm for a channel height of 50 μm, the linear radius-dependence characteristic of the moving speed is still valid (Figure 5d). This clearly indicates the physical origin of e-DEP is electrokinetics (EK) but not electrohydrodynamics (EHD), since EK is usually not sensitive to a finite calculation domain, while EHD is more sensitive to the volume of the computational geometry.

It is worth noting that the speed of movement due to e-DEP at high frequencies (Figure 5b,d) is about one-third of that from ICEP within the low frequency range (Figure 4b,d). Even so, e-DEP may serve as a better method of choice for unidirectional transport of the Janus ME, in that the application

Micromachines **2020**, *11*, 289

of a high field frequency greatly eliminates the unwanted effects of electrochemical polarization and electrode erosion, both of which can cause potential damage to any biological content within the buffer.

3.5. Collection of Nanoscale Analyte Using an Individual Janus Mobile Electrode

On the basis of the fundamental studies in preceding sections, it is then interesting to test whether the direction-controllable AC electrokinetic transport of the Janus ME can be applied for sequential loading and collection of nanoscale biomolecules from the surrounding liquid medium, when it moves continuously in perpendicular orientation to the externally-imposed AC forcing.

Trace molecule assumption is invoked here, namely, the concentration of the biological analyte is orders of magnitude lower than that of the ionic charge carriers (cations and anions) in the electrolyte solution. So, the total charge magnitude of the biomolecules is negligibly small, and its contribution to the charge conservation equation can be safely disregarded. In this sense, all the electric field equations in Section 2 are still valid under current situation. Since an external AC field is employed, the effect of linear electrophoresis time-averages to zero and can then be discarded. The dielectrophoretic (DEP) force of a nanoscale object is also trivial, as evidenced by the quantitative computation shown in Figure 6a. The analyte velocity induced by DEP force field is far below 1 μm/s, albeit the field gradient attains the order of O (10^4–10^5) V/m.

According to the analysis above, the concentration field of the biomolecules is calculated via a traditional convection-diffusion equation for dilute analyte:

$$\frac{\partial c}{\partial t} + \nabla \cdot J = 0 \quad J = u_f c - D\nabla c \tag{15}$$

where c is the concentration field of the charged biomolecules, J the volumetric flux density of the analyte, and D the analyte diffusivity of the secondary particle (SP).

The nanoscale biomolecules around the primary particle (PP) of the Janus ME is supposed to have a diameter of $d = 2r = 100$ nm, which is about two to three orders of magnitude smaller than the Janus ME, so they can be treated as secondary particles (SP). The thermal diffusivity of the biomolecule is 4.3×10^{-12} m^2/s, according to the Einstein relation $D = k_B T/6\eta\pi r$. Here, k_B is the Boltzmann constant, and $T = 293.15$ K the ambient temperature. Impenetrable condition is imposed on the opposing electrode plates with zero normal mass flux. The left and right ends of the chamber are set as open boundaries. Early on, the biological SP is supposed to distribute uniformly within the chamber with a background concentration value of 1 nM.

With the above settings in the simulation software, the transient behavior of loading and collection of the biological SP is then investigated in terms of two distinct situations, including the low-frequency ICEP (Figures 7 and 8) and high-frequency e-DEP delivery of the asymmetric PP (Figures 6 and 9), respectively.

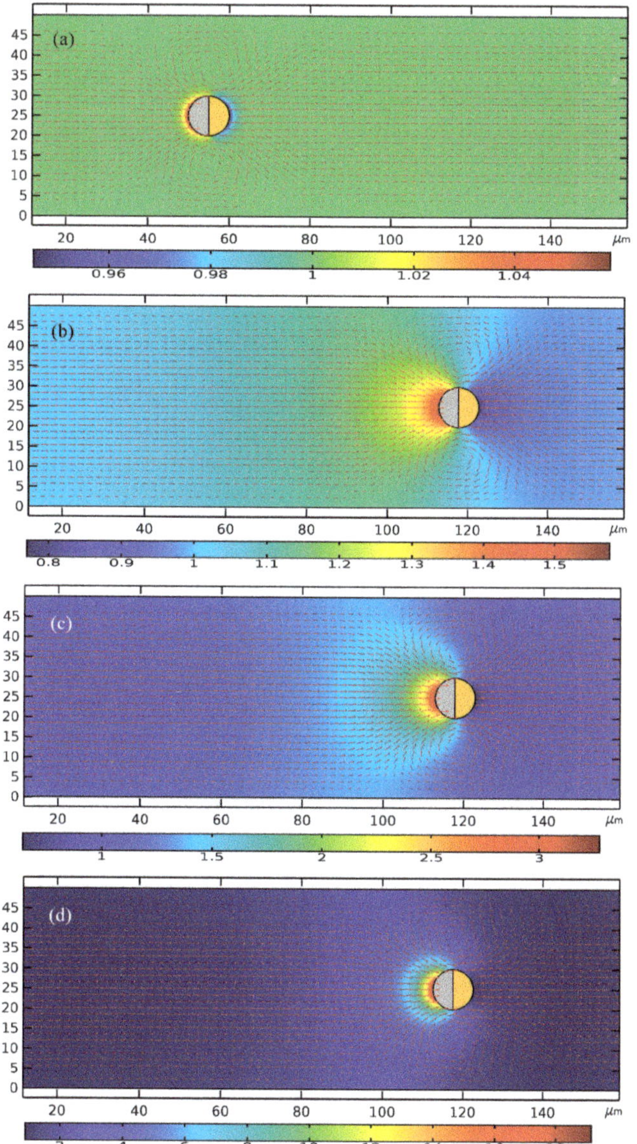

Figure 6. A simulation investigation of the application of the e-DEP-enabled unidirectional delivery of the Janus ME (PP) in simultaneous hydrodynamic loading and collection of the nanoscale biomolecules (SP) from the surrounding liquid suspension in the high-frequency limit. (**a–d**) An arrow plot of the hydrodynamic flow field due to the e-DEP motion of the Janus ME, and a surface plot of the concentration field of the biomolecules (unit: nM) inside the fluidic chamber at $f = 10$ MHz, (**a**) at $t = 0$ s, (**b**) at $t = 2.7$ s for $V_{AC} = 1$ V, (**c**) at $t = 1.2$ s for $V_{AC} = 1.5$ V, and (**d**) at $t = 0.67$ s for $V_{AC} = 2$ V (See Supporting Video S4). In (**b–d**), the Janus PP has moved a same horizontal distance of 60 μm as compared to (**a**). Besides, with increasing voltage from (**b**) to (**d**), the concentrating factor of the SP increases, while the collection area reduces.

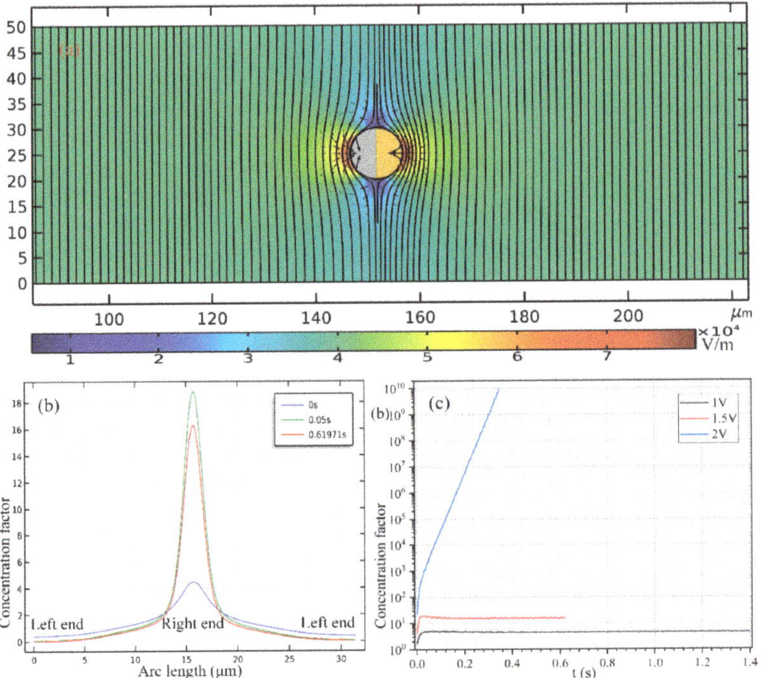

Figure 7. (a) The positive DEP velocity field of the nanoscale biomolecules dispersed in the liquid suspension around the Janus ME having an almost symmetric induced dipole moment in the low-frequency limit, which is on the order of 10^{-10}–10^{-7} m/s and can thereby be ignored in comparison with the electroconvective flow effect. (b) The concentrating factor along the surface of the Janus PP at distinct time instants once the sinewave generator with an AC voltage of 1.5 V at f = 100 Hz is switched on. (c) The peak concentration factor of the target biomolecules on the surface of the Janus ME as a function of time for varying AC voltages at f = 100 Hz.

3.5.1. Loading and Concentrating of SP by a Janus PP in Low Frequency Range

As shown in Figure 8, to enable a vivid visual clarification of the process of cargo loading, we present a surface plot of the SP's concentration field and an arrow plot of the flow field within the liquid domain, along with the ICEP-driven delivery of the mobile PP in the direction of the insulating end towards the entrance. The Janus PP is released at x = 245 μm in the vicinity of the channel exit with a zero initial velocity (Figure 8a). On application of a low-frequency AC signal with a voltage amplitude from 1 V (Figure 8b), 1.5 V (Figure 8c) to 2 V (Figure 8d), ICEO vortex flow field is induced by the applied field preferentially on the surface of the conducing end (Figure 8a). The ME moves in the direction of the insulating end by ICEO propulsion on the membrane surface. As it transports via ICEP towards the entrance, the strong electrohydrodynamic slipping flow field of two oppositely rotating micro-vortices entrain the surrounding biomolecules, and make these SP collect on the rightmost end of the conducting membrane. In fact, this physical process is dynamic in nature, wherein the directed motion of PP and the collection of SP on some portion of the PP surface occur simultaneously, so that the same transient solver is always employed for the calculation of electric-field-mediated fluid-structure interaction and analyte mass transfer in the medium domain.

For the purpose of comparison, we plot the concentration field of SP and the position of PP at a same location x = 175 μm of the PP. Not only much less time is required for the Janus ME to cover an identical distance of 70 μm (245 − 175 = 70 μm), but the peak concentration of the SP collected by

ICEO slipping on the conducting surface increases sharply as well with an enhancement of the applied voltage (Figure 8b–d).

In Figure 7b, we present the concentration profile of the biomolecules on the surface of Janus cylinder at three distinct time instants on switching the function generator on for a voltage of 2 V, including $t = 0$ s, 0.05 s and 0.61971 s. As we can see, with time elapses, the peak concentration increases at first, then begins to decrease at a critical time node, this varying trend is also clearly shown in Figure 7c. Even so, the width of the collection band shrinks monotonously with time (Figure 7b). In addition, though a smaller voltage induces a lower localized concentration factor of the biomolecules due to a reduced ICEO slipping flow field (Figure 7c), a voltage of 1 V gives rise to a larger collection area (Figure 7b) as compared to the case of both 1 V (Figure 8c) and 2 V (Figure 8d). This suggests us to make use of a moderate voltage amplitude for loading and collection of SP in real applications, so we can make a delicate trade-off between the practically realizable trapping area and the concentration factor.

3.5.2. Loading and Concentrating of SP by a Janus PP in High-Frequency Limit

Since the high-frequency e-DEP motion reverses direction with respect to the low-frequency ICEP translation, the Janus PP is placed at x = 55 μm adjacent to the channel entrance before turning on the sinewave generator for high-frequency actuations.

Similar simulation results are presented in Figure 6 for the case of e-DEP as Figure 6 for the situation of ICEP, albeit the initial position of the mobile PP is adjacent to the entrance (Figure 6a) rather than the exit (Figure 6a). Under different voltage amplitudes but a same displacement of 60 μm, the concentration field of the biological SP and the electrokinetic flow field are compared in Figure 6b for 1 V, Figure 6c for 1.5 V and Figure 6d for 2 V. It can be seen that, an increasing voltage implies a higher electrokinetic flow rate caused by a larger particle e-DEP velocity, as well as a higher peak concentration ratio on the surface of the insulating hemisphere (Figure 6). The area of collection of the secondary nanoparticles diminishes as the voltage enhances (Figure 6), nevertheless, which is quite similar to the case of high-frequency excitation (Figure 8).

The time-dependent concentrating factor is provided in Figure 9b. The high-frequency growth trend, however, now manifests as a monotonous increasing trend once ignoring the localized secondary fluctuation (Figure 9b), which forms a stark contrast with the non-monotonous time-dependent variation in the low-frequency case (Figure 7d). In addition, the quantitative data of the concentration value of the biomolecules along the ME's surface in Figure 9c also indicates that, the concentrating efficiency increases, while the area of the concentration belt decreases as the powering time advances.

The most notable difference between the two limiting cases of low- and high-frequency biomolecule loading is that: the swirling motion of the SP around the mobile PP is caused by ICEO whirlpools, which is EHD in essence (Figure 8); on the contrary, the main force that drives the nanoparticles to the surface of the insulating end is the fluid flow from e-DEP motion of the ME, which is of an electrokinetic origin, and this passive electrokinetic flow driven by high-frequency e-DEP (Figure 6) is much slower than the active electrohydrodynamic flow due to low-frequency ICEO slipping (Figure 8). As a result, the concentrating factor is much higher at low frequencies (Figure 7c), in which the active ICEO slipping flow around the constantly translating Janus ME brings large amounts of biomolecules to the conducting end from surrounding electrolyte (Figure 8d).

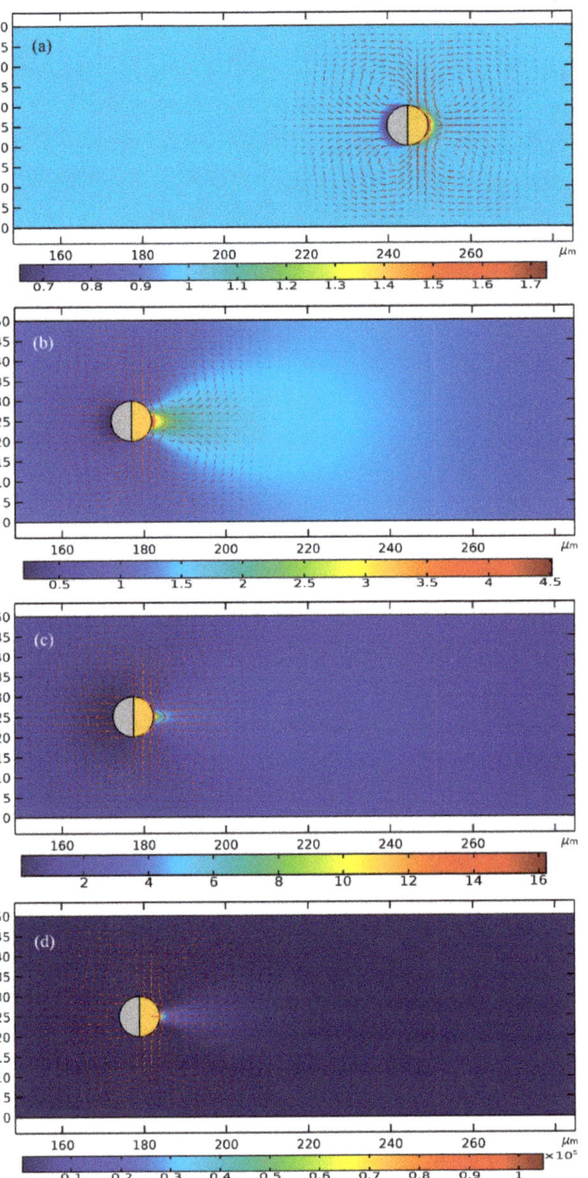

Figure 8. A simulation study of the application of the ICEP-enabled unidirectional delivery of the Janus ME (PP) in simultaneous electroconvective loading and collection of the nanoscale biomolecules (SP) from the surrounding liquid suspension in the low-frequency limit. (**a–d**) An arrow plot of the EHD vortex flow field around the ME, and a surface plot of the concentration field of the biomolecules (unit: nM) inside the fluidic chamber at f = 100 Hz, (**a**) at t = 0 s, (**b**) at t = 0.5 s for V_{AC} = 1 V, (**c**) at t = 0.22 s for V_{AC} = 1.5 V, and (**d**) at t = 0.17 s for V_{AC} = 2 V (See Supporting Video S3). In (**b–d**), the Janus PP has moved a same horizontal distance of 70 μm as compared to (**a**). Besides, with increasing voltage from (**b**) to (**d**), the concentrating factor of the SP enhances sharply, while the collection area shrinks gradually.

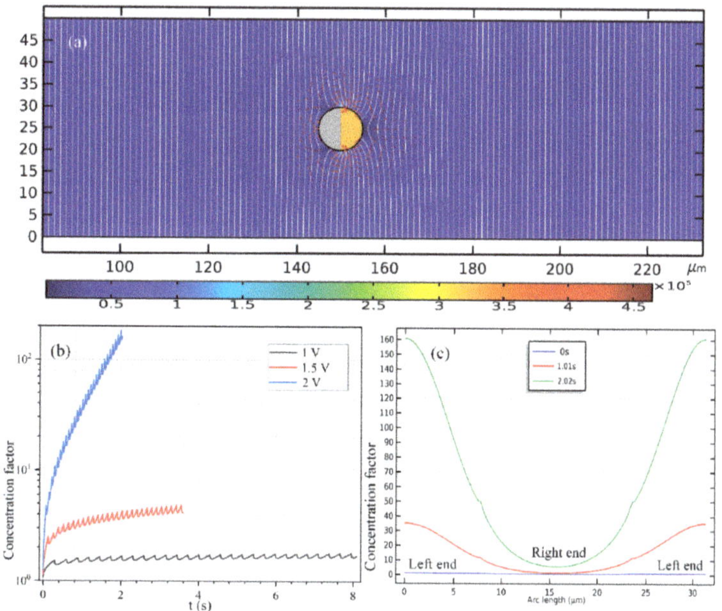

Figure 9. (**a**) The positive DEP velocity field of the nanoscale biomolecules dispersed in the liquid suspension around the Janus ME having an asymmetric induced dipole moment in the high-frequency limit, which is on the order of 10^{-10}–10^{-7} m/s and can thereby be ignored in comparison with the hydrodynamic flow effect. (**b**) The peak concentration factor of the target biomolecules on the surface of the Janus ME as a function of time for varying AC voltages at $f = 10$ MHz, and the translating speed of the PP equals 23.46 μm/s for 1 V, 52.9 μm/s for 1.5 V, and 94 μm/s for 1.5 V. (**c**) The concentrating factor along the surface of the Janus PP at distinct time instants once the sinewave generator with an AC voltage of 1.5 V at $f = 10$ MHz is switched on.

3.6. Continuous Collection of Biomolecules in Buffer Medium using an Array of Janus Mobile Microelectrodes

We then arbitrarily raise the liquid conductivity from the previous 0.001 S/m to 0.1 S/m to approach a more realistic biological culture condition. With such high ionic strength, ICEP decays almost to zero due to ion overcrowding inside the IDL. In this way, the Janus ME can only be delivered by the mechanism of e-DEP, which itself is not essentially limited by the steric effect and still work well at suitable excitation field frequencies. The enhancement of the solution conductivity leads to an augment of the inverse Debye relaxation time of the liquid bulk of $f_{MW} = 22.5$ MHz. According to previous analysis, the imposed field frequency has to be lifted up to exceed this particular value in order for e-DEP to occur. In this sense, a sinusoidal voltage signal at $f = 500$ MHz and $V_0 = 1.5$ V is imposed to the oppositely-polarized driving electrode pair for causing observable e-DEP-induced unidirectional motion of an array of Janus ME with 4 PP placed sequentially along the channel centerline (See Supporting Video S5). To accommodate the increase in the size of ME array, the channel length is correspondingly enlarged from 300 μm to 600 μm (Figure 10a–c).

Micromachines **2020**, *11*, 289

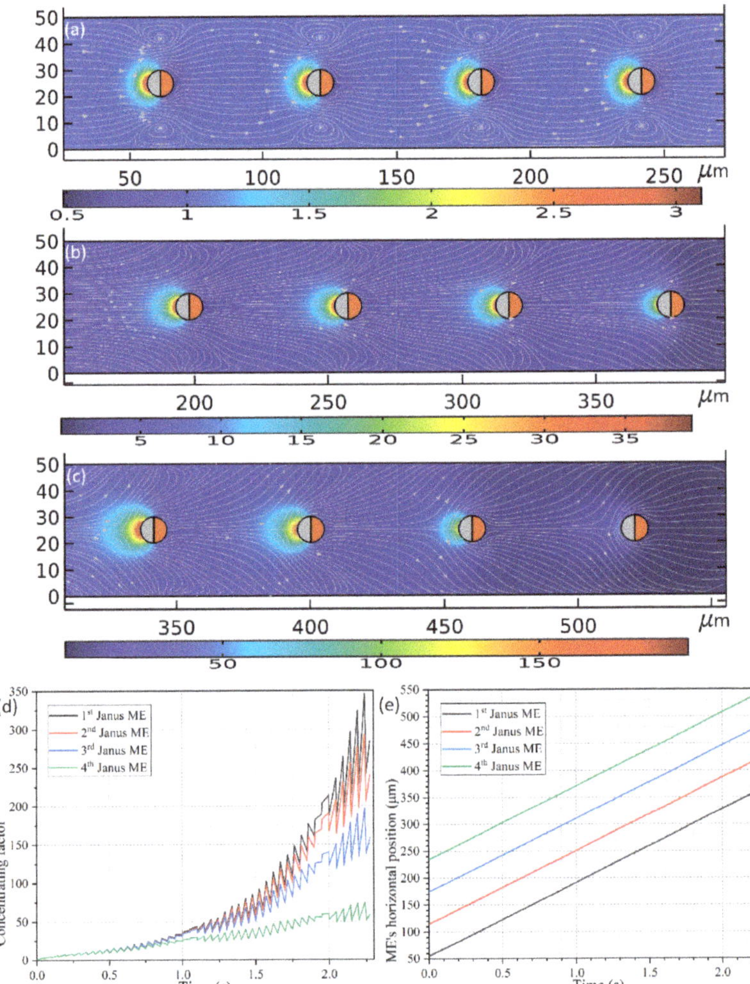

Figure 10. A simulation study of the feasibility of an array of Janus ME with four motile PP arranged along the channel centerline driven by e-DEP force in continuous collection of secondary biomolecules from a 0.1 S/m buffer medium with an AC voltage signal of 1.5 V and 500 MHz. (**a–c**) An arrowed streamline plot of the hydrodynamic flow field around the array of the Janus ME experiencing e-DEP force, and a surface plot of the concentration field of the biomolecules (unit: nM) inside the fluidic chamber, at (**a**) $t = 0.05$ s, (**b**) $t = 1.05$ s, (**c**) $t = 2.1$ s (See Supporting Video S5). (**d**) Temporal evolution of the concentrating factor of the target biomolecules on the surface of the sequential PP as the ME array moves via e-DEP motion. (**e**) Time-dependent horizontal location of the four motile electrodes as they transport unidirectionally towards downstream, from which a time-averaged e-DEP translational velocity of 136.318 μm/s is obtainable.

As shown in Figure 10a, four spherical PP with the line connecting their centers being parallel to the channel length direction are released around the inlet. All the PPs have an identical diameter of 10 μm as used in Section 3.5, and the distance between the nearest ends of adjacent ME is $L_G = 50$ μm. On application of an external harmonic AC forcing of $f = 500$ MHz and 1.5 V to the conducting plates, the four Janus ME move unidirectionally towards the channel outlet at an almost equal translating speed of 136.318 μm/s due to the action of e-DEP mechanism, which is much quicker than that of

52.9 μm/s utilizing an individual ME. The reason behind this acceleration can be vividly accounted for by a making a direct comparison between Figures 6 and 10a–c. As shown in Figure 10a–c, the hydrodynamic convection induced by each Janus ME is in the same direction, and thereby can be effectively superimposed to some extent, which has closing bearing on the interparticle separation (not shown). An enhanced electroconvection helps to transport the ME array much more effectively under the influence of e-DEP, and simultaneously gives rise to a much higher concentrating factor, which is even beyond 100-fold (Figure 10d) in comparison with four-fold induced by a single motile electrode (the red line in Figure 9b).

A fact of interest is that, the peaking concentrating performance evolves differently for the four Janus ME (Figure 10d). Early on, right after switching the functional generator on, the value of concentrating index of the biomolecules grows consistently for all the motile electrodes before $t = 0.75$ s. With the advance of time, the SP concentration factor attains 25-fold on the surfaces of the ME array. Nonetheless, since the PP in the upstream section has a higher chance to capture the dispersed SP, the growth trend of the concentrating efficiency diverges from now on, and more nanoparticles would be trapped on the mobile electrodes they pass through first. For such, although an increasing trend is found for all the Janus ME, the peak concentrating factor decreases step by step as the biomolecules travel from the 1st ME to the 4th ME (Figure 10d). And this phenomenon is most evident in the latest stage as displayed in Figure 10c, implying it serves as a potential method for constructing concentration gradient generators for various on-chip biomedical applications.

In this particular case, the gap size $L_G = 50$ μm between the mobile electrodes keeps constant while they move by e-DEP. In preliminary testing, however, we also simulated the cases in which the interelectrode separation $L_G = 10$–40 μm, and we found that there is active electrokinetic and electroconvective interactions among them, which behaves as a rather complex nature. So, we do not focus on this subject in the current analysis, but can still roughly conclude that, if the interelectrode gap is no less than 10 times the ME's radius, i.e., $L_G \geq 10 \times R$, the kinematic interaction between neighboring Janus entities will be negligibly small and thereby the gap between the electrodes will be maintained while moving them.

3.7. Scaling Analysis of the Stable ICEP Motion In Low-Frequency Limit

An approximate analytical solution of the scale of the induced zeta potential is given by:

$$\zeta \approx \frac{E_B \cdot R}{(1 + \delta)(1 + j\omega\tau_{RC})} \tag{16}$$

where E_B denotes the applied electric field strength, and R the radius of ME, δ the surface capacitance ratio, ω the angular field frequency, $\tau_{RC} = C_D R/\sigma_f(1 + \delta)$ the RC time constant for capacitive charging of the induced double layer.

Conventionally, ICEP insists well in a so-called low-voltage limit, in which the voltage drop across the diffuse part of the induced double layer (IDL) is no more than about a threshold value of $\zeta_0 = 1.5$ V. Note that δ is usually far less than 1, and thereby we have the scaling expression of AC voltage amplitude in DC limit:

$$V_{AC} \leq \frac{\zeta_0 H_C}{R} \tag{17}$$

For the geometry size of $H_C = 50$ μm and $R = 6$ μm used in current analysis, the AC voltage magnitude should be no more than 12.5 V, that is, $V_{AC} \leq 12.5$ V according to the preceding equation, which is in good agreement with the range of V_{AC} ($V_{AC} \leq 4$ V) applied in the parametric study of current work. Under such circumstances, the non-uniform surface electrokinetic transport of counterions within the Debye screening layer becomes quite limited, so do the resulted ion concentration polarization (ICP) phenomenon and electroosmosis of second kind that behaves as a strong electrokinetic eddy in the depleted region, due to a sufficiently small Dukhin number (a physical expression that indicates the ratio of the inhomogeneous surface conduction effect within the EDL to the uniform electrolyte bulk conductivity).

4. Conclusions

In short summary, we have provided results in terms of both mechanism analysis and numerical simulation, to account for the phenomenon of the directional-controllable unidirectional transport of a Janus mobile electrode (ME) free from buoyancy in a uniform background electric field within a broad frequency range, which is from below the inverse RC time constant for the capacitive charging of the IDL on the conducting end to even far beyond the Debye relaxation frequency of the bulk electrolyte. In stark contrast with the ICEP translation at low field frequencies, a new physical phenomenon, named "ego-dielectrophoresis (e-DEP)" is found to exist in the high frequency range, wherein the IDL has already been short circuited by a huge displacement current flowing across the Debye layer. In this way, e-DEP-enabled motion of the Janus colloid is of a pure electrokinetic origin, which behaves specifically as asymmetric interfacial charge relaxation in a homogenous electric field.

Our work justifies for the first time that a buoyancy-free Janus ME or an array of such can be transported unidirectionally in the direction of the conducting end in the high-frequency limit due to the action of driving force from e-DEP, which is right against the movement of ICEP in low frequency range. The scaling characteristic of the translating speed of e-DEP due to localized field perturbation abides by $U_{e\text{-}DEP} \propto V^2 R$, which differs from the traditional DEP velocity scale of a homogenous particle in an externally-imposed field gradient $U_{DEP} \propto V^2 R^2$. So, in practice, it is not necessary for us to arbitrarily increase the size of the Janus ME for achieving a quicker e-DEP-enabled translating speed. Meanwhile, the phenomenon of e-DEP poses a lower demand on the electric conductivity of the liquid suspension as compared to ICEP which can only exists in dilute electrolyte.

The bidirectional moving behavior of the Janus mobile electrode in a wide frequency range is directly pertinent to the handling of nanoscale biological contents in microfluidic channels. The most salient feature of this is its robust dual-functionality in unidirectional delivery of the primary particle (Janus ME) and electroconvective collection of the secondary particles (biomolecules) on the surface of an array of the Janus mobile electrode at the same time. Future research efforts will focus on the practical experimental observation of Janus ME-based electrohydrodynamic microfluidic chips, which exceeds the scope of current analysis. It is believed that the AC electrokinetic behavior of buoyancy-free Janus mobile electrode would actively boost the interdisciplinary research on analytical chemistry, on-chip bioassay, and particle-particle electrokinetic interaction in modern microfluidic systems.

Supplementary Materials: The following are available online at http://www.mdpi.com/2072-666X/11/3/289/s1, Video S1: Rotation and translation of the Janus ME by ICEP at 100 Hz. Video S2: Rotation and translation of the Janus ME by eDEP at 10 MHz. Video S3: Loading of SP by the mobile PP driven by ICEO streaming at 100 Hz. Video S4: Loading of SP by the mobile PP driven by eDEP-enabled flow at 10 MHz. Video S5: Biomolecule concentration using an array of Janus mobile electrode.

Author Contributions: Writing—original draft preparation, W.L.; supervision and project administration, Y.R.; methodology, Y.T.; conceptualization, H.Y.; software, C.X.; Literature retrieval, Q.W. All authors have read and agreed to the published version of the manuscript.

Funding: This project is financially supported by the National Natural Science Foundation of China (No. 11702035, No. 11672284, No. 11702075), Natural Science Foundation of Shaanxi Province (No. 2019JQ-073), Henan provincial department of transportation science and technology project (No. 2019G-2-5), National or Provincial College Students' Innovation and Entrepreneurship Training Program (No. 201910710069), Shaanxi key industrial innovation chain (group)-industrial field (2019ZDLGY15-04-02), and Self-Planned Task (SKLRS201803B) of State Key Laboratory of Robotics and System (HIT).

Conflicts of Interest: The authors declare no conflict of interest.

References

1. Salari, A.; Navi, M.; Lijnse, T.; Dalton, C. AC electrothermal effect in microfluidics: A review. *Micromachines* **2019**, *10*, 762. [CrossRef] [PubMed]

2. Hu, G.; Li, D. Multiscale phenomena in microfluidics and nanofluidics. *Chem. Eng. Sci.* **2007**, *62*, 3443–3454. [CrossRef]

3. García-Sánchez, P.; Ferney, M.; Ren, Y.; Ramos, A. Actuation of co-flowing electrolytes in a microfluidic system by microelectrode arrays. *Microfluid. Nanofluid.* **2012**, *13*, 441–449. [CrossRef]

4. Ren, Y.; Jiang, H.; Yang, H.; Ramos, A.; García-Sánchez, P. Electrical manipulation of electrolytes with conductivity gradients in microsystems. *J. Electrost.* **2009**, *67*, 372–376. [CrossRef]

5. Kunti, G.; Bhattacharya, A.; Chakraborty, S. Analysis of micromixing of non-newtonian fluids driven by alternating current electrothermal flow. *J. Non Newton. Fluid Mech.* **2017**, *247*, 123–131. [CrossRef]

6. Lian, M.; Islam, N.; Wu, J. Ac electrothermal manipulation of conductive fluids and particles for lab-chip applications. *IET Nanobiotechnol.* **2007**, *1*, 36–43. [CrossRef]

7. García-Sánchez, P.; Ren, Y.; Arcenegui, J.J.; Morgan, H.; Ramos, A. Alternating current electrokinetic properties of gold-coated microspheres. *Langmuir* **2012**, *28*, 13861–13870. [CrossRef]

8. Ren, Y.K.; Morganti, D.; Jiang, H.Y.; Ramos, A.; Morgan, H. Electrorotation of metallic microspheres. *Langmuir* **2011**, *27*, 2128–2131. [CrossRef]

9. Liu, Z.; Li, D.; Song, Y.; Pan, X.; Li, D.; Xuan, X. Surface-Conduction enhanced dielectrophoretic-like particle migration in electric-field driven fluid flow through a straight rectangular microchannel. *Phys. Fluids* **2017**, *29*, 102001. [CrossRef]

10. Ramos, A. *Electrokinetics and Electrohydrodynamics in Microsystems*; Springer Science & Business Media: Berlin, Germany, 2011; Volume 530.

11. Ory, S.; Ehud, Y. The Taylor-Melcher leaky dielectric model as a macroscale electrokinetic description. *J. Fluid Mech.* **2015**, *773*, 1–33.

12. Chen, L.; Ma, J.; Guan, Y. Study of an electroosmotic pump for liquid delivery and its application in capillary column liquid chromatography. *J. Chromatogr. A* **2004**, *1028*, 219–226. [CrossRef]

13. Beni, G.; Tenan, M. Dynamics of electrowetting displays. *J. Appl. Phys.* **1981**, *52*, 6011–6015. [CrossRef]

14. Gimsa, J.; Eppmann, P.; Prüger, B. Introducing phase analysis light scattering for dielectric characterization: Measurement of traveling-wave pumping. *Biophys. J.* **1997**, *73*, 3309. [CrossRef]

15. Atten, P. Electrohydrodynamic instability and motion induced by injected space charge in insulating liquids. *IEEE Trans. Dielectr. Electr. Insul.* **1996**, *3*, 1–17. [CrossRef]

16. Feng, Y.; Seyed-Yagoobi, J. Understanding of electrohydrodynamic conduction pumping phenomenon. *Phys. Fluids* **2004**, *16*, 2432–2441. [CrossRef]

17. Squires, T.M.; Bazant, M.Z. Induced-Charge electro-osmosis. *J. Fluid Mech.* **2004**, *509*, 217–252. [CrossRef]

18. Mavrogiannis, N.; Desmond, M.; Gagnon, Z.R. Fluidic dielectrophoresis: The polarization and displacement of electrical liquid interfaces. *Electrophoresis* **2015**, *36*, 1386–1395. [CrossRef]

19. Ding, Z.; Wong, T.N. Electrohydrodynamic instability in an annular liquid layer with radial conductivity gradients. *Phys. Rev. E Stat. Nonlinear Soft Matter Phys.* **2014**, *89*, 033010. [CrossRef]

20. Mavrogiannis, N.; Desmond, M.; Ling, K.; Fu, X.; Gagnon, Z. Microfluidic mixing and analog on-chip concentration control using fluidic dielectrophoresis. *Micromachines* **2016**, *7*, 214. [CrossRef]

21. Bazant, M.Z.; Thornton, K.; Ajdari, A. Diffuse-Charge dynamics in electrochemical systems. *Phys. Rev. E* **2004**, *70*, 021506. [CrossRef]

22. Bonnefont, A.; Argoul, F.; Bazant, M.Z. Analysis of diffuse-layer effects on time-dependent interfacial kinetics. *J. Electroanal. Chem.* **2001**, *500*, 52–61. [CrossRef]

23. Kilic, M.S.; Bazant, M.Z.; Ajdari, A. Steric effects in the dynamics of electrolytes at large applied voltages. I. Double-Layer charging. *Phys. Rev. E* **2007**, *75*, 021502. [CrossRef]

24. Kilic, M.S.; Bazant, M.Z.; Ajdari, A. Steric effects in the dynamics of electrolytes at large applied voltages. Ii. Modified poisson-nernst-planck equations. *Phys. Rev. E* **2007**, *75*, 021503. [CrossRef]

25. Williams, S. AC dielectrophoresis lab-on-chip devices. In *Encyclopedia of Microfluidics and Nanofluidics*; Springer: Berlin, Germany, 2008; pp. 1–8.

26. Morgan, H.; Green, N.G. *Ac Electrokinetics: Colloids and Nanoparticles*; Research Studies Press: Baldock, UK, 2003.

27. Zhao, C.; Yang, C. Advances in electrokinetics and their applications in micro/nano fluidics. *Microfluid. Nanofluid.* **2012**, *13*, 179–203. [CrossRef]

28. Ristenpart, W.; Aksay, I.; Saville, D. Electrically guided assembly of planar superlattices in binary colloidal suspensions. *Phys. Rev. Lett.* **2003**, *90*, 128303. [CrossRef]

29. Beale, S.C. Capillary electrophoresis. *Anal. Chem.* **1998**, *70*, 279–300. [CrossRef]
30. Boymelgreen, A.; Yossifon, G. Observing electrokinetic janus particle-channel wall interaction using microparticle image velocimetry. *Langmuir* **2015**, *31*, 8243–8250. [CrossRef]
31. Kilic, M.S.; Bazant, M.Z. Induced-Charge electrophoresis near a wall. *Electrophoresis* **2011**, *32*, 614–628. [CrossRef]
32. Rose, K.A.; Hoffman, B.; Saintillan, D.; Shaqfeh, E.S.; Santiago, J.G. Hydrodynamic interactions in metal rodlike-particle suspensions due to induced charge electroosmosis. *Phys. Rev. E* **2009**, *79*, 011402. [CrossRef]
33. Gangwal, S.; Cayre, O.J.; Bazant, M.Z.; Velev, O.D. Induced-Charge electrophoresis of metallodielectric particles. *Phys. Rev. Lett.* **2008**, *100*, 058302. [CrossRef]
34. Zhao, H.; Bau, H.H. On the effect of induced electro-osmosis on a cylindrical particle next to a surface. *Langmuir* **2007**, *23*, 4053–4063. [CrossRef] [PubMed]
35. Saintillan, D.; Darve, E.; Shaqfeh, E.S. Hydrodynamic interactions in the induced-charge electrophoresis of colloidal rod dispersions. *J. Fluid Mech.* **2006**, *563*, 223–259. [CrossRef]
36. Lewpiriyawong, N.; Kandaswamy, K.; Yang, C.; Ivanov, V.; Stocker, R. Microfluidic characterization and continuous separation of cells and particles using conducting poly (dimethyl siloxane) electrode induced alternating current-dielectrophoresis. *Anal. Chem.* **2011**, *83*, 9579–9585. [CrossRef] [PubMed]
37. Urdaneta, M.; Smela, E. Multiple frequency dielectrophoresis. *Electrophoresis* **2007**, *28*, 3145–3155. [CrossRef] [PubMed]
38. Sun, H.; Ren, Y.; Hou, L.; Tao, Y.; Liu, W.; Jiang, T.; Jiang, H. Continuous particle trapping, switching, and sorting utilizing a combination of dielectrophoresis and alternating current electrothermal flow. *Anal. Chem.* **2019**, *91*, 5729–5738. [CrossRef] [PubMed]
39. Yariv, E. "Force-Free" electrophoresis? *Phys. Fluids* **2006**, *18*, 61. [CrossRef]
40. Gascoyne, P.R.; Vykoukal, J. Particle separation by dielectrophoresis. *Electrophoresis* **2002**, *23*, 1973. [CrossRef]
41. Jiang, T.; Ren, Y.; Liu, W.; Tang, D.; Tao, Y.; Xue, R.; Jiang, H. Dielectrophoretic separation with a floating-electrode array embedded in microfabricated fluidic networks. *Phys. Fluids* **2018**, *30*, 112003. [CrossRef]
42. Miloh, T. A unified theory of dipolophoresis for nanoparticles. *Phys. Fluids* **2008**, *20*, 107105. [CrossRef]
43. Rodríguez-Sánchez, L.; Ramos, A.; García-Sánchez, P. Electrorotation of semiconducting microspheres. *Phys. Rev. E* **2019**, *100*, 042616. [CrossRef]
44. García-Sánchez, P.; Ramos, A. Electrorotation and electroorientation of semiconductor nanowires. *Langmuir* **2017**, *33*, 8553–8561. [CrossRef] [PubMed]
45. Daghighi, Y.; Li, D. Micro-Valve using induced-charge electrokinetic motion of janus particle. *Lab Chip* **2011**, *11*, 2929–2940. [CrossRef] [PubMed]
46. Zhang, L.; Zhu, Y. Dielectrophoresis of janus particles under high frequency ac-electric fields. *Appl. Phys. Lett.* **2010**, *96*, 141902. [CrossRef]
47. Gangwal, S.; Pawar, A.; Kretzschmar, I.; Velev, O.D. Programmed assembly of metallodielectric patchy particles in external ac electric fields. *Soft Matter* **2010**, *6*, 1413–1418. [CrossRef]
48. Gangwal, S.; Cayre, O.J.; Velev, O.D. Dielectrophoretic assembly of metallodielectric janus particles in Ac electric fields. *Langmuir* **2008**, *24*, 13312–13320. [CrossRef]
49. Velev, O.D.; Gangwal, S.; Petsev, D.N. Particle-Localized ac and dc manipulation and electrokinetics. *Annu. Rep. Sect. C Phys. Chem.* **2009**, *105*, 213–246. [CrossRef]
50. Ramos, A.; García-Sánchez, P.; Morgan, H. Ac electrokinetics of conducting microparticles: A review. *Curr. Opin. Colloid Interface Sci.* **2016**, *24*, 79–90. [CrossRef]
51. Tang, S.Y.; Zhu, J.; Sivan, V.; Gol, B.; Soffe, R.; Zhang, W.; Mitchell, A.; Khoshmanesh, K. Creation of liquid metal 3d microstructures using dielectrophoresis. *Adv. Funct. Mater.* **2015**, *25*, 4445–4452. [CrossRef]
52. Tang, S.Y.; Sivan, V.; Petersen, P.; Zhang, W.; Morrison, P.D.; Kalantar-Zadeh, K.; Mitchell, A.; Khoshmanesh, K. Liquid metal actuator for inducing chaotic advection. *Adv. Funct. Mater.* **2015**, *24*, 5851–5858. [CrossRef]
53. Tang, S.Y.; Khoshmanesh, K.; Sivan, V.; Petersen, P.; O'Mullane, A.P.; Abbott, D.; Mitchell, A.; Kalantar-Zadeh, K. Liquid metal enabled pump. *Proc. Natl. Acad. Sci. USA* **2014**, *111*, 3304. [CrossRef] [PubMed]
54. So, J.-H.; Dickey, M.D. Inherently aligned microfluidic electrodes composed of liquid metal. *Lab Chip* **2011**, *11*, 905–911. [CrossRef]
55. Dickey, M.D. Stretchable and soft electronics using liquid metals. *Adv. Mater.* **2017**, *29*, 1606425. [CrossRef]

56. Khoshmanesh, K.; Tang, S.-Y.; Zhu, J.Y.; Schaefer, S.; Mitchell, A.; Kalantar-Zadeh, K.; Dickey, M.D. Liquid metal enabled microfluidics. *Lab Chip* **2017**, *17*, 974–993. [CrossRef] [PubMed]
57. Boymelgreen, A.; Yossifon, G.; Miloh, T. Propulsion of active colloids by self-induced field gradients. *Langmuir* **2016**, *32*, 9540–9547. [CrossRef]
58. Boymelgreen, A.M.; Balli, T.; Miloh, T.; Yossifon, G. Active colloids as mobile microelectrodes for unified label-free selective cargo transport. *Nat. Commun.* **2018**, *9*, 1–8. [CrossRef]
59. Wu, Y.; Fu, A.; Yossifon, G. Active particles as mobile microelectrodes for selective bacteria electroporation and transport. *Sci. Adv.* **2020**, *6*, eaay4412. [CrossRef]
60. Loget, G.; Kuhn, A. Bulk synthesis of janus objects and asymmetric patchy particles. *J. Mater. Chem.* **2012**, *22*, 15457–15474. [CrossRef]
61. Kunti, G.; Bhattacharya, A.; Chakraborty, S. Alternating current electrothermal modulated moving contact line dynamics of immiscible binary fluids over patterned surfaces. *Soft Matter* **2017**, *13*, 6377–6389. [CrossRef] [PubMed]
62. Kunti, G.; Bhattacharya, A.; Chakraborty, S. Rapid mixing with high-throughput in a semi-active semi-passive micromixer. *Electrophoresis* **2017**, *38*, 1310–1317. [CrossRef]
63. Stubbe, M.; Gyurova, A.; Gimsa, J. Experimental verification of an equivalent circuit for the characterization of electrothermal micropumps: High pumping velocities induced by the external inductance at driving voltages below 5 v. *Electrophoresis* **2013**, *34*, 562–574. [CrossRef] [PubMed]
64. Gao, J.; Sin, M.L.; Liu, T.; Gau, V.; Liao, J.C.; Wong, P.K. Hybrid electrokinetic manipulation in high-conductivity media. *Lab Chip* **2011**, *11*, 1770–1775. [CrossRef] [PubMed]
65. Wu, J.; Lian, M.; Yang, K. Micropumping of biofluids by alternating current electrothermal effects. *Appl. Phys. Lett.* **2007**, *90*, 234103. [CrossRef]
66. Feldman, H.C.; Sigurdson, M.; Meinhart, C.D. Ac electrothermal enhancement of heterogeneous assays in microfluidics. *Lab Chip* **2007**, *7*, 1553–1559. [CrossRef]
67. Kua, C.H.; Lam, Y.C.; Rodriguez, I.; Yang, C.; Youcef-Toumi, K. Dynamic cell fractionation and transportation using moving dielectrophoresis. *Anal. Chem.* **2007**, *79*, 6975–6987. [CrossRef] [PubMed]
68. Hossan, M.R.; Dutta, D.; Islam, N.; Dutta, P. Review: Electric field driven pumping in microfluidic device. *Electrophoresis* **2018**, *39*, 702–731. [CrossRef] [PubMed]
69. Olesen, L.H.; Bruus, H.; Ajdari, A. Ac electrokinetic micropumps: The effect of geometrical confinement, faradaic current injection, and nonlinear surface capacitance. *Phys. Rev. E* **2006**, *73*, 056313. [CrossRef]
70. Wu, J.J. Ac electro-osmotic micropump by asymmetric electrode polarization. *J. Appl. Phys.* **2008**, *103*, 024907. [CrossRef]
71. Green, N.G.; Ramos, A.; González, A.; Morgan, H.; Castellanos, A. Fluid flow induced by nonuniform ac electric fields in electrolytes on microelectrodes. I. Experimental measurements. *Phys. Rev. E* **2000**, *61*, 4011. [CrossRef]
72. González, A.; Ramos, A.; Green, N.G.; Castellanos, A.; Morgan, H. Fluid flow induced by nonuniform ac electric fields in electrolytes on microelectrodes. Ii. A linear double-layer analysis. *Phys. Rev. E* **2000**, *61*, 4019. [CrossRef]
73. Green, N.G.; Ramos, A.; Gonzalez, A.; Morgan, H.; Castellanos, A. Fluid flow induced by nonuniform ac electric fields in electrolytes on microelectrodes. III. Observation of streamlines and numerical simulation. *Phys. Rev. E* **2002**, *66*, 026305. [CrossRef] [PubMed]
74. Ramos, A.; Morgan, H.; Green, N.G.; Castellanos, A. Ac electrokinetics: A review of forces in microelectrode structures. *J. Phys. D Appl. Phys.* **1998**, *31*, 2338. [CrossRef]
75. González, A.; Ramos, A.; Castellanos, A. Pumping of electrolytes using travelling-wave electro-osmosis: A weakly nonlinear analysis. *Microfluid. Nanofluid.* **2008**, *5*, 507–515. [CrossRef]
76. García-Sánchez, P.; Ramos, A.; González, A.; Green, N.G.; Morgan, H. Flow reversal in traveling-wave electrokinetics: An analysis of forces due to ionic concentration gradients. *Langmuir* **2009**, *25*, 4988–4997. [CrossRef] [PubMed]
77. Cahill, B.P.; Heyderman, L.J.; Gobrecht, J.; Stemmer, A. Electro-Osmotic streaming on application of traveling-wave electric fields. *Phys. Rev. E* **2004**, *70*, 036305. [CrossRef]
78. Thamida, S.K.; Chang, H.-C. Nonlinear electrokinetic ejection and entrainment due to polarization at nearly insulated wedges. *Phys. Fluids* **2002**, *14*, 4315–4328. [CrossRef]

79. Wang, S.-C.; Chen, H.-P.; Lee, C.-Y.; Yu, C.-C.; Chang, H.-C. Ac electro-osmotic mixing induced by non-contact external electrodes. *Biosens. Bioelectron.* **2006**, *22*, 563–567. [CrossRef] [PubMed]
80. Chang, H.-C.; Yossifon, G.; Demekhin, E.A. Nanoscale electrokinetics and microvortices: How microhydrodynamics affects nanofluidic ion flux. *Annu. Rev. Fluid Mech.* **2012**, *44*, 401–426. [CrossRef]
81. Ren, Y.; Liu, W.; Tao, Y.; Hui, M.; Wu, Q. On ac-field-induced nonlinear electroosmosis next to the sharp corner-field-singularity of leaky dielectric blocks and its application in on-chip micro-mixing. *Micromachines* **2018**, *9*, 102. [CrossRef]
82. Bazant, M.Z.; Squires, T.M. Induced-Charge electrokinetic phenomena: Theory and microfluidic applications. *Phys. Rev. Lett.* **2004**, *92*, 066101. [CrossRef]

micromachines

Article

Low-Power, Multimodal Laser Micromachining of Materials for Applications in sub-5 µm Shadow Masks and sub-10 µm Interdigitated Electrodes (IDEs) Fabrication

Cacie Hart [1,2] and Swaminathan Rajaraman [1,2,3,4,*]

1   Department of Materials Science & Engineering, University of Central Florida, Orlando, FL 32816, USA; chart@knights.ucf.edu
2   NanoScience Technology Center, University of Central Florida, Orlando, FL 32816, USA
3   Department of Electrical & Computer Engineering, University of Central Florida, Orlando, FL 32816, USA
4   Burnett School of Biomedical Sciences, University of Central Florida, Orlando, FL 32816, USA
*   Correspondence: Swaminathan.Rajaraman@ucf.edu; Tel.: +1-407-823-4339

Received: 16 January 2020; Accepted: 7 February 2020; Published: 8 February 2020

Abstract: Laser micromachining is a direct write microfabrication technology that has several advantages over traditional micro/nanofabrication techniques. In this paper, we present a comprehensive characterization of a QuikLaze 50ST2 multimodal laser micromachining tool by determining the ablation characteristics of six (6) different materials and demonstrating two applications. Both the thermodynamic theoretical and experimental ablation characteristics of stainless steel (SS) and aluminum are examined at 1064 nm, silicon and polydimethylsiloxane (PDMS) at 532 nm, and Kapton® and polyethylene terephthalate at 355 nm. We found that the experimental data aligned well with the theoretical analysis. Additionally, two applications of this multimodal laser micromachining technology are demonstrated: shadow masking down to approximately 1.5 µm feature sizes and interdigitated electrode (IDE) fabrication down to 7 µm electrode gap width.

Keywords: multimodal laser micromachining; ablation characteristics; shadow mask; interdigitated electrodes

1. Introduction

In recent years, laser technology has been examined as an exciting material processing method for both industrial and academic researchers [1–4]. The high quality of laser beams allows for improved micromachining precision for a variety of materials [5,6]. Additionally, the compact size, high efficiency, cost-effectiveness, direct machining, 3D fabrication, and ease of integration are appealing to academic researchers with limited benchtop area and lean budgets.

Laser micromachining, specifically, involves the ablation of materials where the features produced by the laser are in the micro-scale [6,7]. Laser micromachining techniques are currently employed in the automobile [8], medical [2,9], semiconductor [10,11], and solar cell industries [12]. Lasers used in micromachining are available in a wide range of wavelengths (ultraviolet to infrared), pulse durations (micro- to femtosecond), and repetition rates (single pulse to megahertz) [6,8]. Because of this flexibility, laser micromachining allows for the processing of a variety of materials.

Figure 1 compares laser micromachining to three commonly used direct-write microfabrication methods [13–19]. All these methods involve a mechanism for removing material directly from a substrate in a desired pattern using computer-controlled machinery. While other patterning methods, such as photolithography are often used, they are not direct-write techniques and involve several steps to pattern materials; thus, a comparison to these other methods is not provided in this figure.

Photolithography is, of course, required for one of the methods depicted (reactive ion etching or RIE) for performing feature definition followed by RIE to "engrave' the feature in the substrate beneath [15–17,19]. Since RIE requires photolithography, the process involves more steps when compared to the other methods depicted in Figure 1 [17]. Micromilling is a very technically simple process; however, this simplicity comes at the expense of frequent drill bit brakeage and the inability to produce features in the sub-100 µm range [14]. As with some RIE technologies, focused ion beam (FIB) milling can define nanoscale features; however, the write process can take days depending on the pattern complexity. Additionally, in FIB-based milling processes, the material surface may become amorphous due to the implantation of gallium ions [13,16–18]. The comparison depicted in Figure 1 is by no means comprehensive and has been included to show the characteristics of laser-based micromachining with some other commonly used technologies for patterning materials in micromachining and micro-electro-mechanical systems (MEMS). Several other techniques for direct subtractive fabrication of micromachined features are available such as electrical discharge machining (EDM) [20], ultrasonic drilling [21], water jet cutting [22] etc.

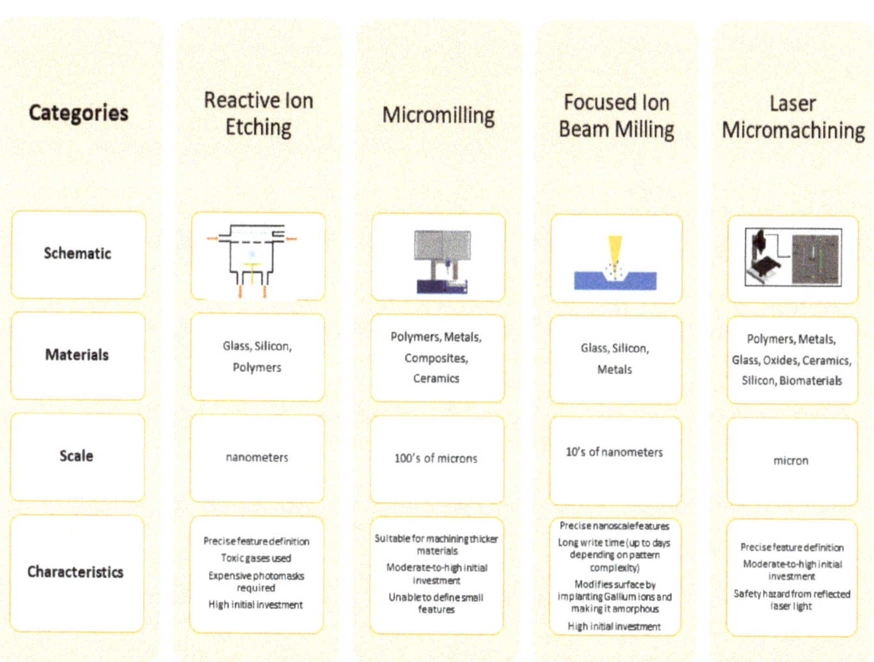

Figure 1. Comparison of three widely used direct-write fabrication methods (reactive ion etching (RIE) [15,17], micromilling [14], and focused ion beam (FIB) [13,16,18]) to laser micromachining. Each of these methods allows for high aspect ratio, serial, single-substrate fabrication. In the case of RIE multi-wafer/multi-substrate parallel processing is also possible as with some laser micromachining examples as well for higher throughput micro/nanofabrication.

Laser micromachining has occasionally been used for shadow mask fabrication; however, the state of the art in laser micromachining (minimum feature size of 10 µm) [23] (to date) has relied entirely on single-wavelength excimer, CO_2, and Nd:YAG (neodymium-doped yttrium aluminum garnet) lasers [1,3,12,23–36]. To the best of our knowledge, multimodal laser micromachining has not been used for shadow mask fabrication. While the term "multimodal" in the laser field typically means that the laser contains multiple transverse electromagnetic modes, the laser micromachining tool used in this work has the ability to operate at several wavelengths in the same tool, a rare capability not afforded by many lasers. Specifically, with respect to shadow mask microfabrication with lasers, several impressive research efforts are reported in literature: Klank, et. al. fabricated 85 µm channels in polymethyl methacrylate (PMMA) with a CO_2 laser [1]. Fan, et. al. also used a CO_2 laser to micromachine 250 µm lines in wax for use as a shadow mask [3]. Tahir et. al. used a 1064 nm Nd:YAG laser to fabricate 250 µm channels in wood, glass, plastic, and rubber [12]. Chung, et. al. fabricated shadow masks with minimum feature sizes of 200 µm with a 785 nm Ti: sapphire laser [26]. Shiu, et. al. machined 140 µm channels in low-carbon steel for use as a shadow mask using an excimer laser [34].

The lasers used in the aforementioned research can produce shadow masks for MEMS applications, but they are unable to micromachine highly precise features on the scale of a single micron, unless they operate in the femtosecond regime [3,7,23,24,30,37]. The higher power of these lasers allows for the processing of thicker materials in a more reasonable time scale; however, this capability comes at the price of benchtop machining, high costs, space, high power usage (GW), and absence of multimodality to micromachine several materials with the same tool. Characterization of lasers used for materials processing is necessary for the proper use of such tools. By understanding the necessary energy, repetition rate/frequency, spot size, and depths that can be micromachined with various processing conditions in specific materials, users can subsequently optimize their experiments for successful results in different applications.

In this paper, we report the full characterization of a multimodal laser micromachining tool and two applications of the usage of microstructures fabricated using the tool. Multimodal laser micromachining allows for a wide range of materials processing capabilities; however, this comes at the price of limited power and nanosecond ablation. These tools allow users to operate at specific laser wavelengths (1064 nm, 532 nm, and 355 nm, in our case) and reduced powers (2.6 mJ maximum power). These conditions, however, provide greater control over the material selectivity for laser micromachining and ablation depths. Specifically, one can remove a material from a device while leaving the remainder of the constituent materials unaffected [38].

Typically, such laser micromachining tools are employed in liquid crystal display (LCD) repair, semiconductor failure analysis, and removing shorts and passivation layers in integrated circuits [6, 39]. In this paper, we characterize the multimodal tool for the laser micromachining of six (6) different materials: Stainless steel (SS) and aluminum, to be machined with infrared (IR) mode, polyethylene terephthalate (PET) and Kapton®, to be machined with ultraviolet (UV) mode, and polydimethylsiloxane (PDMS) and silicon, to be machined with green mode. Subsequent shadow mask patterning allows for the definition of organic and inorganic layers and the development of a fully functional interdigitated electrode (IDE) devices. These devices are further characterized in this paper.

2. Microfabrication Method Overview

In laser micromachining, the laser beam is collimated into a small spot and patterning is achieved by either moving the substrate within a fixed beam or rastering the laser across a surface [6]. The desired machining patterns simply need to be drawn in a CAD program (such as AutoCAD, Autodesk, San Lafayette, CA, USA) and imported as a drawing exchange format (DXF) file into the control program of the laser micromachining tool. Once the program is executed with the laser, substrate material removal can be a result of photochemical, photothermal, or photophysical ablation [40], as shown in Figure 2. Commonly used processes include laser cutting, scribing, drilling, or etching to produce relief structures or holes on a substrate in ambient temperatures [3,8,23,27,40–42]. The power of this technique lies in the ability to construct desired patterns on arbitrarily shaped surfaces, with the only limitation being the degrees of freedom and the resolution of the motion controller. Laser micromachining is considered a rapid prototyping technique because designs can be changed immediately without the need to fabricate new molds or masks.

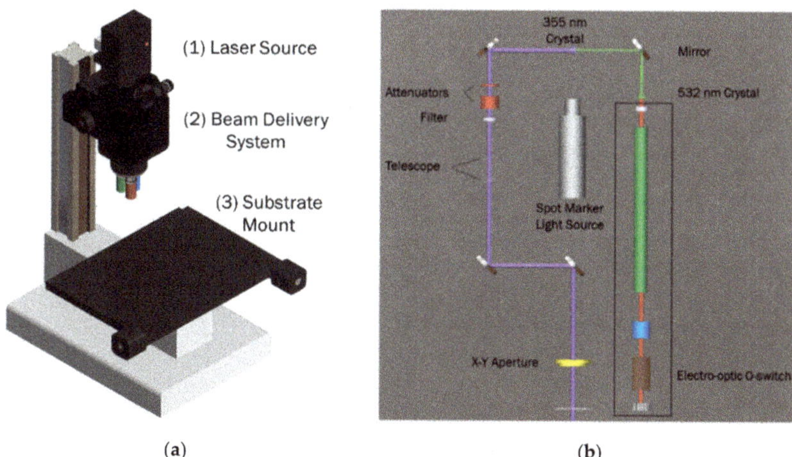

Figure 2. Mechanisms of ablation in laser micromachining. Substrate removal can be a result of thermal ablation, physical ablation, chemical ablation, or a combination of these mechanisms.

Every laser micromachining system is comprised of three parts, as shown in Figure 3a for the multimodal laser micromachining tool: (1) the source laser, (2) the beam delivery system, and (3) the substrate mounting stage. Obviously, the laser source is at the heart of the system, as it determines which substrates and feature sizes can be micromachined [8]. The system used in this work has a multimodal laser source that allows for switching between three wavelengths of light: 1064 nm infrared (IR mode), 532 nm visible green (Visible mode), and 355 nm ultra-violet (UV mode). This wavelength switching allows for greater substrate compatibility and a wide array of feature sizes in an extremely compact benchtop system.

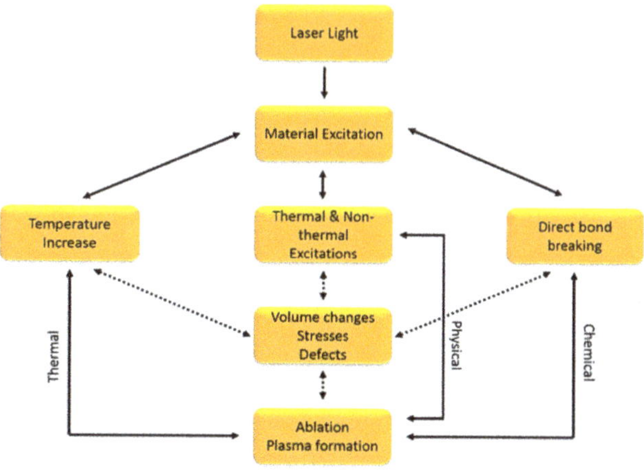

(a)   (b)

Figure 3. (a) External schematic of a QuikLaze 50ST2 Multimodal Laser Micromachining System. The three components of every laser micromachining system are shown: (1) the laser source box; (2) the beam delivery system; and (3) the motorized substrate mount. (b) Internal schematics of the laser source box. This shows the three crystals and a switch that create the different wavelengths of light the system produces with the ability to switch between three wavelengths.

Micromachines **2020**, *11*, 178

Beam delivery involves optical components, including fixed focusing objects and mirrors, galvanometric scanners, optical fibers, wave-guides, apertures, and q-switches, that are used to generate the laser spot. The selection of these optical components depends on the working distance, desired spot size, and required energy [6,8,41,43]. The combination of the laser source and optical components of the beam delivery system determines the ultimate properties of the laser beam. The beam delivery system used in this work is depicted in Figure 3b with greater detail.

Lastly, the substrate mounting system depends on how the rastering occurs on the tool. If the laser beam is to be rastered over the substrate surface, then a stationary substrate mount may be used. However, in most cases, including the system used in this work, it is more desirable to raster the substrate itself within a laser beam. In this instance, the substrate mount is manipulatable in the x- and y-directions, and even in the z-direction in some cases.

In the multimodal laser system used in this work, the laser beam is initially generated as a 1064 nm IR mode, shown in Figure 3b. The IR laser beam is subsequently passed through two crystals in the beam delivery system that produce the two additional wavelengths. The beam delivery system employs filters prior to the microscope optics to filter out the unnecessary wavelengths, allowing only the desired wavelength to interact with the substrate surface.

3. Theoretical Background

A thermodynamic approach to calculating the theoretical "depth of cut" for the materials tested in this work was adapted from Schütz, et. al. [10]. This formula is based solely on material properties and laser energy. Examining the depth of cut in this manner allows for a more fundamental understanding of the interaction between the laser and the material with fewer assumptions when compared to a molecular approach used for photochemical ablation [10,42,44].

$$a_p = \frac{\Phi}{\rho \int_{T_{room}}^{T_v} c_p(T)dT + \sum_{i=1}^{n} H_{ph}^v} - d \tag{1}$$

Equation (1) shows the relationship between the depth of cut per pulse (a_p), the laser fluence (Φ), and involved material properties, including density (ρ), vaporization temperature (T_v), heat capacity (c_p), and the phase change enthalpy (H^v_{ph}). The d term (right-hand side of the equation) is a correction term that includes optical and thermal losses. This equation represented schematically (Figure 4) balances intrinsic energy and enthalpy, which are state variables and omits fundamental process parameters that are not state variables (i.e., particle dynamics) [45,46]. While the inclusion of these fundamental process parameters could lead to a more accurate theoretical solution, the complexity would be vastly increased due to the use of non-linear partial differential equations and the need for sophisticated simulation software to solve such equations [5,6,10,40,44,47]. Additionally, since this laser micromachining setup does not operate in the femtosecond regime, ablation occurs through melt expulsion and redeposition driven by the vapor pressure and the recoil pressure of light [44,47,48]. As a result, a simple thermodynamic analysis was performed to extract the relationship between the "depth of cut" and laser fluence which can be compared to the results obtained through experiments.

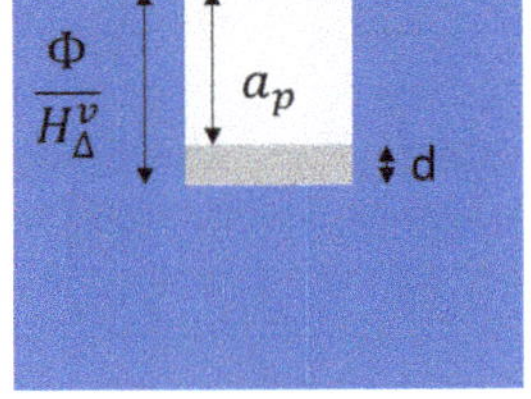

Figure 4. Schematic showing the variables in the thermodynamic theory for evaluation of the depth of cut.

4. Materials and Methods

4.1. Multimodal Laser

A QuikLaze 50ST2 multimodal laser (New Wave Research Inc., Fremont, CA, USA) was used for all the laser micromachining performed in this paper. A selection of three wavelengths as mentioned in the previous sections allows the laser to be tailored to a specific application. The microscope of the laser system is equipped with 10×, 50×, and 100× lenses, each with specific wavelength limitations. Because of additional filters in the microscope lenses, the green wavelength can be used through any of the lenses, while the UV and the IR can only be used through the 50× and 100× lenses, respectively.

The laser outputs a 5 mm diameter Gaussian beam, which is then shaped into a rectangle by the XY aperture. The size of this rectangle is determined by user inputs into the control software. The maximum pulse duration of the laser is 5 ns for all wavelengths; however, it can be adjusted by the user in the program. Additionally, the laser fluence depends on the user specifications in the control software and the wavelength of light used, as the laser output energy can be adjusted by the user. The fluence ranges from a maximum of 27,000 J/cm^2 to a minimum of 1.08 J/cm^2.

4.2. Materials Used

To ensure a comprehensive study of multimodal laser micromachining, several materials were machined. Success in laser micromachining for a given material is typically determined only by the choice of wavelength because each material reacts differently to a specific wavelength. In general, metals absorb shorter wavelengths more effectively than longer wavelengths [6,8]; however, there are limitations to how effectively material is removed at shorter wavelengths determined by the microscope optics used. Because the amount of light transmitted through the microscope objectives required for each wavelength varies, the "effective" absorption of the material at a given wavelength is changed.

Stainless steel 12.5 µm thick (type 304) (Trinity Brand Industries, Countryside, IL, USA) and 16 µm-thick aluminum foil (Reynolds Group Holdings, Auckland, NZ, USA) were machined using the 1064 nm IR wavelength through the 100× microscope lens. The absorbances of these materials at this wavelength is 37% and 5%, respectively [42]. Kapton® of thickness 12.5 µm (DuPont, Wilmington, DE, USA) and 25 µm-thick polyethylene terephthalate (PET) (McMaster-Carr, Elmhurst, IL, USA) were machined using the 355 nm UV wavelength through the 50× microscope lens. The absorbances of these materials at this wavelength are 22.5% and 12%, respectively [40,43]. Finally, Silicon (University Wafer, Boston, MA, USA) and polydimethylsiloxane (PDMS) (Dow Corning, Midland, MI, USA) were micromachined using the 532 nm green wavelength through the 10× microscope lens. The absorbances for these materials at this wavelength are 25% and 72%, respectively [10,12,49].

4.3. Laser Characterization

To establish protocols for processing each material, characterization grids were machined for all six (6) materials. These grids consisted of 100 spots, as shown schematically in Figure 5, and were designed in SolidWorks (Dassault Systems, Waltham, MA, USA).

DXF pattern files were subsequently uploaded into the New Wave Laser program, and each spot was assigned a specific frequency (range of 5 Hz to 50 Hz) increasing by 5 Hz increments along the x-axis of the grid, and a specific energy from 10% (0.27 mJ) to 100% (2.7 mJ) increasing along the negative y-axis of the grid by 10%. The laser spot size was varied for each grid in order to provide full characterization of the laser's capabilities over a wide range of fluence. Grids were subsequently patterned in all 6 materials using the multimodal laser.

All grids were imaged using scanning electron microscopy (JEOL JSM-6480, Tokyo, Japan). Full images of the grids were obtained, as well as images of the individual spots in both flat and 45° angled orientations. ImageJ (NIH, Bethesda, MD, USA) was used to characterize the depth of the laser cut, as well as the resultant spot size.

		Frequency (Hz)									
		5	10	15	20	25	30	35	40	45	50
Energy (mJ)	0.27	∎	∎	∎	∎	∎	∎	∎	∎	∎	∎
	0.54	∎	∎	∎	∎	∎	∎	∎	∎	∎	∎
	0.81	∎	∎	∎	∎	∎	∎	∎	∎	∎	∎
	1.08	∎	∎	∎	∎	∎	∎	∎	∎	∎	∎
	1.35	∎	∎	∎	∎	∎	∎	∎	∎	∎	∎
	1.62	∎	∎	∎	∎	∎	∎	∎	∎	∎	∎
	1.89	∎	∎	∎	∎	∎	∎	∎	∎	∎	∎
	2.16	∎	∎	∎	∎	∎	∎	∎	∎	∎	∎
	2.43	∎	∎	∎	∎	∎	∎	∎	∎	∎	∎
	2.70	∎	∎	∎	∎	∎	∎	∎	∎	∎	∎

Figure 5. Layout of the grid of spots laser cut into the materials. The grid was designed to have 100 μm distances between the spots.

4.4. Shadow Masks

Shadow masks for the patterning of materials were fabricated from Kapton®, SS, PET, and aluminum substrates. These materials were chosen because they are all commonly available in most microfabrication laboratories and it was determined that they could be ablated all the way through the material using the multimodal laser, a necessity for shadow masks. Full coverage, circle-on-line IDE shadow masks were designed using SolidWorks and machined using the multimodal laser and the appropriate wavelength for the substrate material. The shadow masks were then imaged using the scanning electron microscope (SEM, JEOL JSM-6480, Tokyo, Japan) to characterize the design (CAD dimensions) to device (fabricated shadow mask) translation. These shadow masks were subsequently used to pattern both metal and gelatin for design to device studies, as described in Section 4.5.

Additionally, traditional interwoven comb IDE shadow masks were fabricated to test the lowest feature size limits of the multimodal laser. These structures allow for more rapid shadow mask fabrication at the lowest possible widths ensuring higher sensitivity in IDE assays; thus, allowing for any necessary adjustments to achieve the desired electrode gap width. Due to the size of these structures, atomic force microscopy (AFM) (Anasys Instruments, Santa Barbara, CA, USA) and SEM (JEOL, Tokyo, Japan) were used to characterize the electrode gap widths of these shadow masks.

Lastly, because of the widespread use of PDMS in microfluidics, several lines of 2 mm length were laser micromachined into this material in order to determine the depth of cut for a given number of laser passes. All lines were micromachined through the IR microscope lens with the green laser at 2.7 mJ with an X-Y aperture area of 50 μm^2. The number of laser passes was varied from 5 to 40 passes. In this mode of operation, the laser beam is continuously scanned across the surface of the material in the desired pattern.

The process flow for the shadow mask fabrication and subsequent patterning of organic and inorganic layers is depicted in Figure 6.

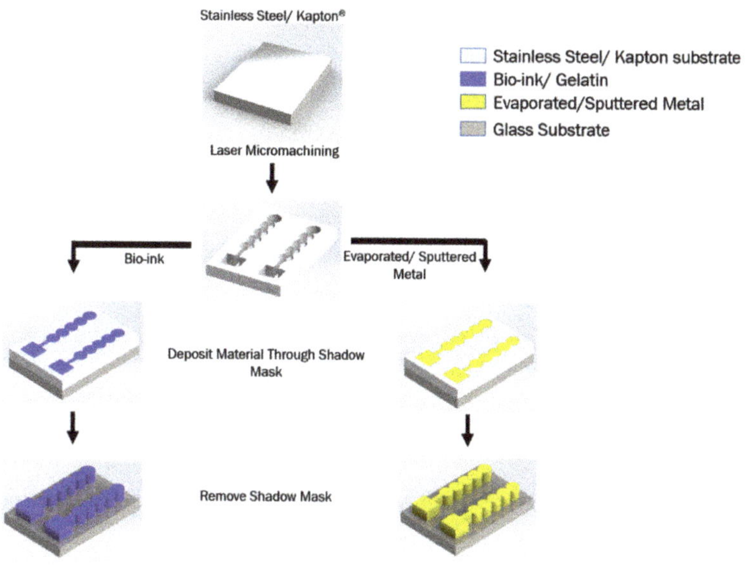

Figure 6. Schematic of laser micromachined shadow mask fabrication and subsequent materials patterning demonstration.

4.5. Patterning of Organic and Inorganic Layers

In order to access the accuracy of low power, multimodal laser micromachining, material patterning through multimodal laser micromachined shadow masks was performed.

4.5.1. Metal Patterning

The Kapton®, SS, aluminum, and PET multimodal laser micromachined shadow masks were affixed onto glass microscope slides (Fisher Scientific, Hampton, NH) with Kapton® tape (DuPont, Wilmington, DE, USA). Titanium-Gold (Ted Pella, Redding, CA, USA) electrodes, traces, and contact pads of 5nm-30 nm thickness respectively were subsequently deposited (Deposition rate: 1 Å/s at 1×10^{-6} Torr) through these shadow masks onto glass microscope slides via electron beam evaporation (Thermionics, Port Townsend, WA, USA). After metal patterning, the shadow masks were removed by carefully detaching the Kapton® tape from the glass slide with tweezers. Transmitted light microscopy (TS2 Inverted Microscope, Nikon, Tokyo, Japan) was used to image the patterned metal structures. ImageJ (NIH, Bethesda, MD) was used for further optical analysis of the captured images and measurement of the electrode structures.

4.5.2. Gelatin Patterning

Gelatin (Millipore Sigma, St. Louis, MO, USA) and deionized (DI) water were mixed to make a 5 wt % gelatin solution to act as a sample bioink. Kapton® IDE shadow masks were dipped in DI water to allow for better adherence to the glass microscope slide substrates. The 5 wt % gelatin solution was subsequently pipetted onto the glass slides through the shadow masks. The gelatin solution was allowed to set, then the shadow masks were carefully removed to expose the gelatin IDE patterns. The resultant gelatin structures were imaged using a Nikon TS2 inverted microscope.

4.6. Impedance Characterization of Metal Patterns

The full spectrum (10 Hz to 100 kHz) impedance of each of the resultant patterned metal IDE structures was measured with a BODE 100 impedance measurement system (Omicron Labs, Klaus,

Austria). Dulbecco's phosphate buffered saline (DPBS) (Gibco, Waltham, MA, USA) acted as the electrolyte solution and platinum-titanium wire (eDAQ, Deniston East, NSW, Australia) was used as a counter electrode used during the impedance measurements. Impedance data was extracted and plotted with Origin 2016 (OriginLab, North Hampton, MA, USA).

5. Results and Discussion

5.1. Laser Characterization

Laser characterization grids were fabricated for each of the six materials for a range of spot sizes as described in the Materials and Methods section. The spot sizes for the materials depended on the microscope objective that was used for the ablation. The spot sizes ranged from 50 µm down to 2 µm for IR ablation through the 100× lens for SS and aluminum, 60 µm down to 4 µm for UV ablation through the 50× lens for Kapton® and PET, and 250 µm down to 20 µm for green ablation though the 10× lens for silicon and PDMS. SEM imaging of these grid structures, and subsequent processing in ImageJ was used to calculate the depth of cut for each spot micromachined by the laser. Figure 7 shows examples of the characterization grids for each of the six materials at the maximum spot size for each wavelength. For PET, Kapton®, SS, and aluminum, full ablation (through vias, all the way through the substrate) is achieved for the full range of power and frequency combinations shown in Figure 5. Since these materials were fully ablated within the 10 pulses used, the ablation depth was averaged over the number of spots. Some deviation from the theoretical calculation for these materials was observed. Ablation rates of 46.3 µm/mJ for Kapton® and SS, 92.6 µm/mJ for PET, and 59.3 µm/mJ for aluminum were achieved. Neither silicon nor PDMS were able to be fully ablated through for any power and frequency combination. Silicon began measurable ablation at a minimum of 35 Hz and 2.43 mJ and had an ablation rate of 1.4 µm/mJ. PDMS began measurable ablation at a minimum of 35 Hz and 2.16 mJ, which gives an ablation rate of 1.5 µm/mJ.

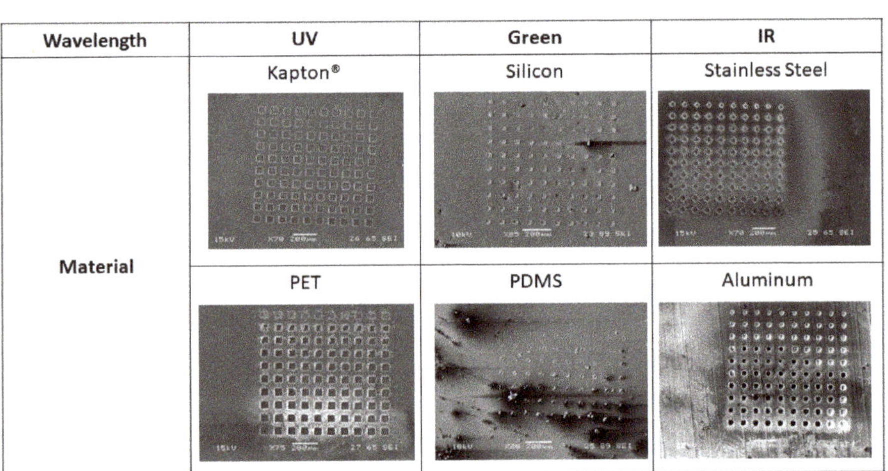

Wavelength	UV	Green	IR
Material	Kapton®	Silicon	Stainless Steel
	PET	PDMS	Aluminum

Figure 7. Scanning electron microscope (SEM) images of the laser characterization grids for each of the six materials. Kapton®, SS, polyethylene terephthalate (PET), and aluminum are all at the maximum spot size for their respective wavelengths. Silicon and polydimethylsiloxane (PDMS) are both at 50% spot size to show the grid in its entirety. All scale bars are 200 µm.

Traces (lines) were scribed in PDMS using the green laser through the IR lens with between 5 and 40 passes and a 50 µm laser spot size. Cross-sectioning the scribed lines carefully using a razor blade, followed by SEM imaging of the cross section, resulted in the measurement of the depth of cut for each number of laser passes. Figure 8 shows the linear relationship that was found between the depth of cut and the number of laser micromachining passes. Characterizing this relationship allows for the fabrication of microchannels of specific depths in PDMS by varying the number of laser passes.

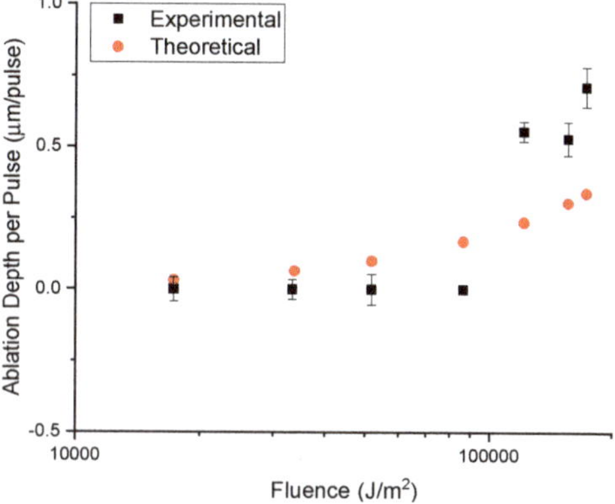

Figure 8. Results from scribing lines in PDMS with a varied number of laser passes. The depth of cut increases linearly with the number of laser passes as expected. This allows for the direct tuning of microchannel depth by selecting the number of passes the laser makes in order to cut PDMS.

Comparison to Theoretical Values

Figure 9 and Table 1 show the comparison between the theoretical ablation depth model and our experimental results. For silicon and PDMS, the thermodynamic model fits well. The comparison in ablation depth per pulse for both materials between theory and practice is in the same order of magnitude. Furthermore, the ablation depth per pulse/ depth of cut plots for both materials show a clear correlation between the experimental and theoretical data. Neither one of these materials are ablated all the way through by the laser.

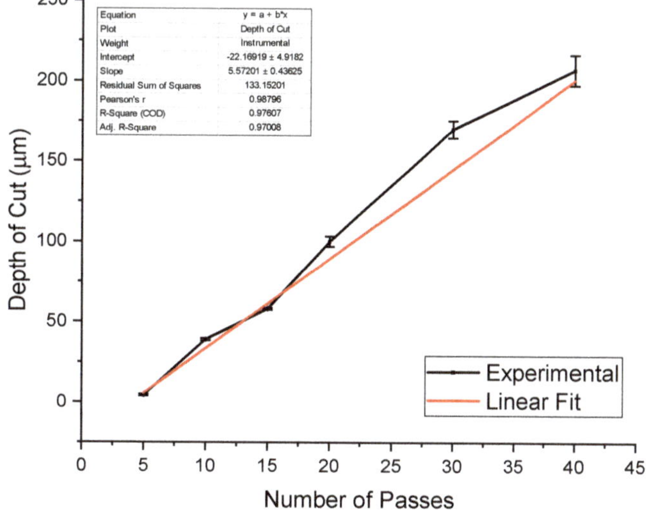

Figure 9. Comparison of theoretical and experimental ablation depth for a range of fluences in silicon.

Table 1. Comparison of theoretical and experimental ablation depths.

Material	Thickness (μm)	Maximum Theoretical Ablation Depth Per Pulse (μm/pulse)	Experimental Ablation Depth per Pulse (μm/pulse)	Ablated Through After 10 Pulses?
Kapton	12.5	2.65	1.25 *	Yes
Stainless Steel	12.5	3.18	1.25 *	Yes
PET (polyethylene terephthalate)	25	15.6	2.5 *	Yes
Aluminum	16.3	16.3	1.63 *	Yes
Silicon	500	0.34024	0.7112	No
PDMS (polydimethylsiloxane)	100	0.25783	0.6481	No

* Limited due to full ablation of materials in less than 10 pulses.

Conversely, deviation from theory is observed in the experimental ablation depth per pulse comparison for Kapton®, SS, PET, and aluminum. Kapton® and SS perform similar to silicon and PDMS as observed in Table 1 with both the theoretical and experimental values in the same order of magnitude with experimental values being smaller than theoretical values. The discrepancy between PET and aluminum is roughly an order of magnitude higher predicted theoretical values. The laser ablated through these materials at some point (difficult to measure experimentally in our current setup), so deviation is expected in all these materials. For example, both Kapton® and SS could have shown an average ablation depth of ~3 μm per pulse, and both PET and aluminum could have shown an average ablation depth of approximately ~16 μm per pulse. Because of the thickness of these materials, when the normalization of the ablation depth to the number of pulses is performed, the experimental values are determined to be substantially lower. The key results from the experimental data in this case is that the laser is capable of fully ablating these thicknesses and that a reasonable number of pulses can be used to ablate even thicker samples of these materials. These results also show the limitations of a thermodynamic approach, as the material thickness is not considered in the calculations.

5.2. Applications

Two applications of this multimodal laser micromachining technique were additionally demonstrated in this work namely the microfabrication of shadow masks and IDEs.

5.2.1. Shadow Masks

Shadow masking technology is an integral part of fabricating micro/nanostructures for prototyping in microelectronics, optical, microfluidic, MEMS, packaging, and biomedical lab-on-a-chip applications [11,23]. Typical methods for producing shadow masks, such as photolithography and deep reactive ion etching (DRIE) or ion beam milling, are expensive, require cleanroom-based fabrication, expensive vacuum equipment, ultra-pure air filtration, and advanced know-how [3,50]. Multimodal laser micromachining, on the other hand, is simple, cost effective, and makerspace-compatible, all vital attributes for cell-based assays and microfluidics applications.

Fabrication of the shadow masks down to ~1.5 μm was successfully demonstrated, as shown in Figure 10. To the best of our knowledge, this is the lowest feature size demonstrated for laser defined shadow masks [3,23]. Previous work with laser micromachining has produced feature sizes down to ~10 μm. As a result, patterns that are an almost an order of magnitude better than the state of the art are reported in this paper.

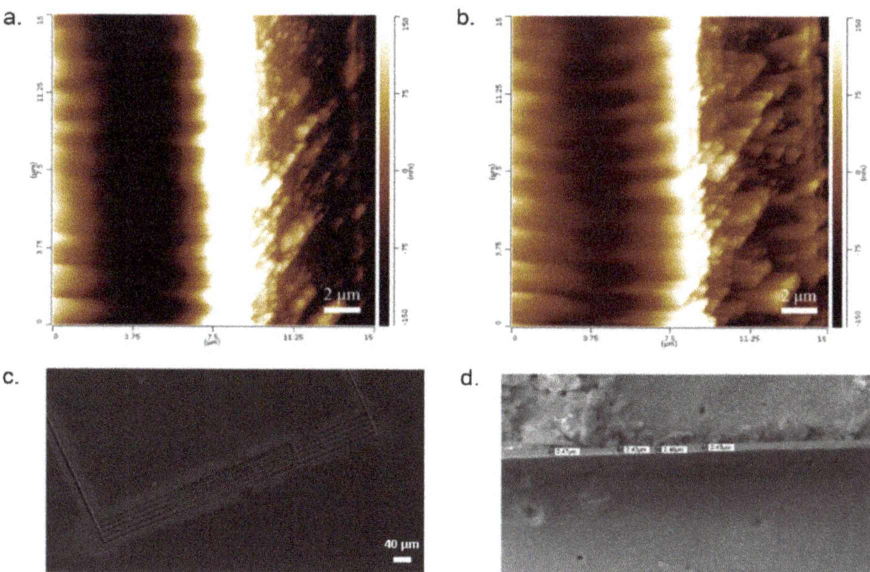

Figure 10. Images of sub-5 µm trace widths in Kapton®. (**a**) Atomic force microscopy (AFM) image of 3.5 µm trace width (white area). (**b**) AFM image of 1.5 µm trace width (white area). (**c**) SEM image of full comb finger electrode structure (approx. 5 µm trace width). (**d**) SEM image of shadow mask feature of approximately 2.43 µm trace width.

5.2.2. Patterning Through Shadow Masks

The laser micromachined shadow masks were further used to pattern both metal and gelatin/bio-ink as described in Section 4.5. The organic and inorganic layer patterning can be utilized for applications such as accurate cell placement in single cell and culture assays, precision confinement and growth of cellular constructs, tissue engineering, metal micro/nanoelectrodes, definition of organic insulation layers, and other lab-on-a-chip and diagnostic applications [3,23,41,49,51].

Design of a shadow mask to microfabricated device translation for the four materials that the multimodal laser was completely able to micromachine in its entirety are shown in Figure 11. Aluminum, PET and SS shadow masks were used to fabricate metal IDEs while Kapton shadow masks were used for fabricating a gelatin/bio-ink IDE. It was observed that Kapton® and SS demonstrated the best design to device translation for 125 µm to 7 µm. Both SS and Kapton® showed minimal thermal damage from laser microfabrication at their respective laser wavelengths (1064 nm and 355 nm). Additionally, both materials have coefficients of thermal expansion (both approximately 20×10^{-6} K^{-1}) which are 2× larger than the thermal expansion of glass (9×10^{-6} K^{-1}) [46] theoretically suggesting better translation results and experimentally verified in Figure 11 (for N = 3 measurements at the various design values). The best design to device translation was observed to be ~98% for the 7 µm gelatin features due to the rapid deposition and curing of the gelatin. The worst design to device translation for SS was observed to be 95% for 125 µm metal IDE on glass due to the higher run-off possibilities during the electron beam evaporation process [5]. PET and aluminum demonstrated a deviation from the designed IDE pitch by 11.2% and 25.8% at maximum pitch, respectively. PET has a coefficient of thermal expansion (80×10^{-6} K^{-1}) [46] that is nearly an order of magnitude larger than that of glass, so some deviation from the design dimensions is expected due to thermal mismatch effects during e-beam evaporation. The aluminum shadow mask exhibited physical melting during the laser micromachining process, which could explain the large deviation from design dimensions.

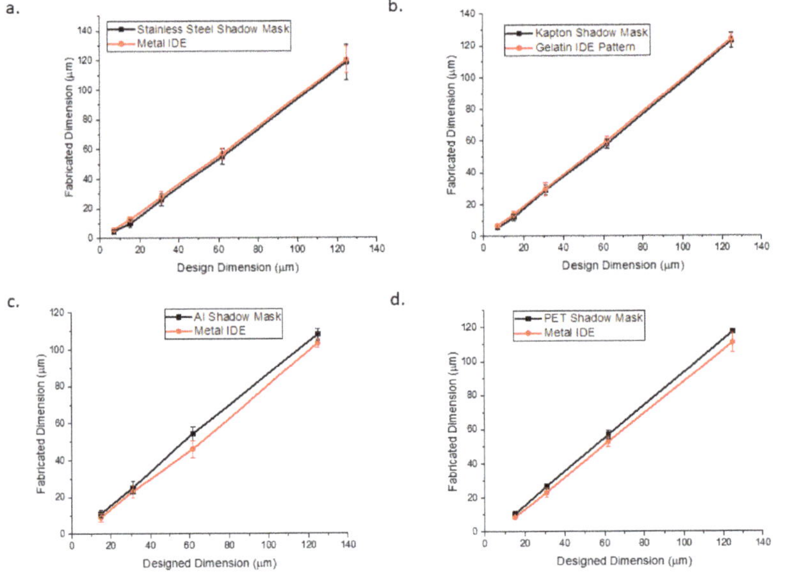

Figure 11. Design to shadow mask to device (interdigitated electrode (IDE) on glass) for the four materials that cut all the way through the substrate: (**a**) stainless steel to Ti-Au metal IDE; (**b**) Kapton® to gelatin IDE pattern; (**c**) aluminum to Ti-Au metal IDE, and (**d**) PET to Ti-Au metal IDE.

Figure 12 shows the design schematic, SEM images of the shadow masks, and transmitted light microscope images of both Ti-Au metal patterning and gelatin bioink patterning for circular fill IDE patterns with electrode gaps from 125 µm to 7 µm. Metal patterning appeared to work best with Kapton® or SS shadow masks because their coefficients of thermal expansion are closer to that of glass. Kapton® and PET shadow masks worked best for gelatin patterning. Both shadow mask materials allowed for simple adhesion to the glass substrate utilizing surface tension effects by simply dipping the mask in DI water prior to attachment on glass substrates. Allowing the gelatin to fully set prior to shadow mask removal was key in preventing the patterns from bleeding.

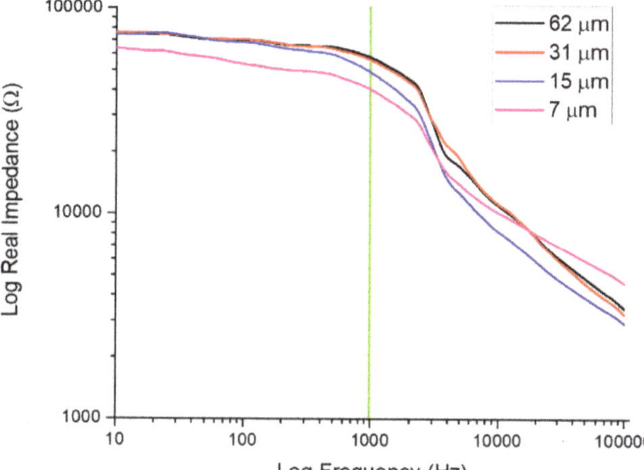

Figure 12. Results of laser micromachining of the shadow masks, as well as inorganic and organic layer patterning. The design of an interdigitated electrode (IDE) was translated into metal (titanium-gold) and bioink (gelatin).

The impedance of the metallized IDEs (Figure 13) was found to decrease with decreasing electrode gap width as is expected [52] with the 1 kHz impedance decreasing by 33.47% between the extremities of the designs tested. Table 2 further illustrates impedances at key frequencies, clearly depicting resistive behavior at the lower and upper ends of the spectra and capacitive behavior in the mid-band with increasing impedance values as expected [52].

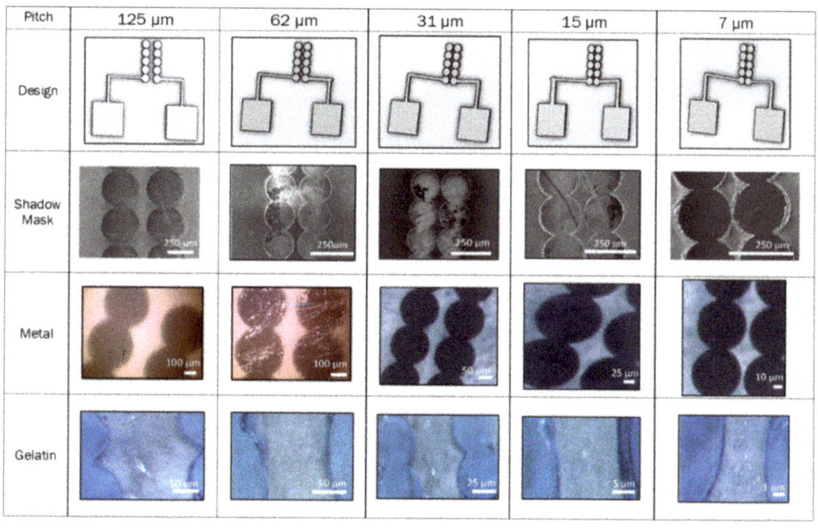

Figure 13. Full spectrum impedance measurements of several IDEs of varying pitch. The electrophysiologically significant frequency of 1 kHz (green line) reports impedances from approximately 39 kΩ (7 μm) to 56 kΩ (62 μm).

Table 2. Values of the real part of the impedance at significant frequencies.

Impedance at Significant Frequencies	Frequency	Real Impedance (kΩ)				
		10 Hz	100 Hz	1 kHz	10 kHz	100 kHz
	7	63.76	53.50	40.40	10.17	4.60
Electrode Gap (µm)	15	74.21	68.37	48.91	8.24	2.90
	32	76.06	70.29	56.06	11.19	3.26
	62	74.77	70.05	57.74	10.99	3.46
		Resistive	Capacitiv	Capacitiv	Resistive	Resistive

6. Conclusions

Complete characterization of the laser micromachining processes for six (6) commonly used microfabrication materials was developed in this work using a multimodal laser micromachining tool. Characterization of the QuikLaze 50ST2 multimodal laser for the laser micromachining of six (6) different materials demonstrated that the ablation depths that were experimentally obtained fit relatively well with a simple thermodynamic theory for most of the materials. While more complex theories or analysis [47] could improve the discrepancy, this technique is accurate enough to allow one to readily calculate possible ablation depths of a new material using such a laser micromachining tool. Additionally, two applications of multimodal laser micromachining were demonstrated: shadow mask fabrication and patterning of organic and inorganic materials in the sub-5 µm range and IDE fabrication in the sub-10 µm range. The ability of such a technique to allow for rapid prototyping of shadow masks and devices, combined with the compact, benchtop-friendly design gives multimodal laser micromachining tremendous promise as an efficient fabrication method in academic and industrial research settings.

Author Contributions: C.H. and S.R. conceived and designed the experiments. C.H. developed the theoretical background for laser micromachining. She additionally performed all the laser micromachining experiments with the six (6) different materials. C.H. imaged all the samples and plotted the data. C.H. and S.R. analyzed the data and interpreted it. C.H. and S.R. wrote and edited the paper. All authors have read and agreed to the published version of the manuscript.

Funding: This research received no external funding. It was supported by Rajaraman's startup funding at UCF.

Acknowledgments: The authors would like to thank Charles Didier for helping design the schematics. We would also like to thank the UCF Materials Characterization Facility for use of the SEM. Additionally, we would like to thank the Rajaraman's startup funding at UCF that funded this work.

Conflicts of Interest: The authors declare no conflict of interest.

References

1. Klank, H.; Kutter, J.P.; Geschke, O. CO2-laser micromachining and back-end processing for rapid production of PMMA-based microfluidic systems. *Lab Chip* **2002**, *2*, 242–246. [CrossRef]
2. Anderson, R.R.; Margolis, R.J.; Watenabe, S.; Flotte, T.; Hruza, G.J.; Dover, J.S. Selective photothermolysis of cutaneous pigmentation by q-switched Nd:YAG laser pulses at 1064, 532, and 355 nm. *J. Investig. Dermatol.* **1989**, *93*, 28–32. [CrossRef] [PubMed]
3. Fan, Y.; Li, H.; Yi, Y.; Foulds, I.G. Laser micromachined wax-covered plastic paper as both sputter deposition shadow masks and deep-ultraviolet patterning masks for polymethylmethacrylate-based microfluidic systems. *J. Micro/Nanolithography MEMS MOEMS* **2013**, *12*, 049701. [CrossRef]
4. Hung, Y.H.; Chien, H.L.; Lee, Y.C. Excimer laser three-dimensional micromachining based on image projection and the optical diffraction effect. *Appl. Sci.* **2018**, *8*, 1690. [CrossRef]
5. Arnold, M.S.B.A.C.B. Fundamentals of Laser-Material Interaction and Application to Multiscale Surface Modification. In *Laser Precision Microfabrication*; Sugioka, M.M.K., Pique, A., Eds.; Springer: Berlin/Heidelberg, Germany, 2010.
6. Schaeffer, R.D. *Fundamentals of Laser Micromachining*; CRC Press: Boca Raton, FL, USA, 2012.

7. Rizvi, N.H.; Apte, P. Developments in laser micro-machining techniques. *J. Mater. Process. Technol.* **2002**, *127*, 206–210. [CrossRef]
8. Arnold, C.B.; Piqué, A. Laser Direct-Write Processing. *MRS Bull.* **2007**, *32*, 9–15. [CrossRef]
9. Miller, P.R.; Aggarwal, R.; Doraiswamy, A.; Lin, Y.J.; Lee, Y.S.; Narayan, R.J. Laser micromachining for biomedical applications. *JOM* **2009**, *61*, 35–40. [CrossRef]
10. Schütz, V.; Stute, U.; Horn, A. Thermodynamic investigations on the laser ablation rate of silicon over five fluence decades. *Phys. Procedia* **2013**, *41*, 640–649. [CrossRef]
11. Gower, M.C. Laser micromachining for manufacturing MEMS devices. In *Micromachining and Microfabrication*; SPIE: Bellingham, WA, USA, 2001; Volume 4559.
12. Tahir, B.A.; Ahmed, R.; Ashiq, M.G.B.; Ahmed, A.; Saeed, M.A. Cutting of nonmetallic materials using Nd:YAG laser beam. *Chin. Phys. B* **2012**, *21*, 044201. [CrossRef]
13. Petit, D.; Faulkner, C.C.; Johnstone, S.; Wood, D.; Cowburn, R.P. Nanometer scale patterning using focused ion beam milling. *Rev. Sci. Instrum.* **2005**, *76*, 026105. [CrossRef]
14. Guckenberger, D.J.; de Groot, T.E.; Wan, A.M.; Beebe, D.J.; Young, E.W. Micromilling: A method for ultra-rapid prototyping of plastic microfluidic devices. *Lab Chip* **2015**, *15*, 2364. [CrossRef] [PubMed]
15. Jansen, H.; Gardeniers, H.; de Boer, M.; Elwenspoek, M.; Fluitman, J. A survey on the reactive ion etching of silicon in microtechnology. *J. Micromech. Microeng.* **1995**, *6*, 14. [CrossRef]
16. Utke, I.; Hoffmann, P.; Melngailis, J. Gas-assisted focused electron beam and ion beam processing and fabrication. *J. Vac. Sci. Technol. B* **2008**, *26*, 1197. [CrossRef]
17. Karouta, F. A practical approach to reactive ion etching. *J. Phys. D Appl. Phys.* **2014**, *47*, 233501. [CrossRef]
18. Tseng, A.A. Recent developments in micromilling using focused ion beam technology. *J. Micromech. Microeng.* **2004**, *14*, R15. [CrossRef]
19. Peake, G.M.; Zhang, L.; Li, N.Y.; Sarangan, A.M.; Willison, C.G.; Shul, R.J.; Hersee, S.D. A micromachined, shadow-mask technology for the OMVPE fabrication of integrated optical structures. *J. Electron. Mater.* **2000**, *29*, 86–90. [CrossRef]
20. Jamwal, A.; Aggarwal, A.; Gautam, N.; Devarapalli, A. Nishant Gautam, Akhil Devarapalli, Electro-Discharge Machining: Recent Developments and Trends. *Int. Res. J. Eng. Technol.* **2018**, *5*, 433–448.
21. Kumar, S.; Dvivedi, A. On machining of hard and brittle materials using rotary too micro-ultrasonic drilling process. *Mater. Manuf. Process.* **2019**, *34*, 736–748. [CrossRef]
22. Liu, X.; Liang, Z.; Wen, G.; Yuan, X. Waterjet Machining and Research Developments: A Review. *Int. J. Adv. Manuf. Technol.* **2019**, *102*, 1257–1335. [CrossRef]
23. Kumar, A.; Gupta, A.; Kant, R.; Akhtar, S.N.; Tiwari, N.; Ramkumar, J.; Bhattacharya, S. Optimization of laser machining process for the preparation of photomasks and its application to microsystems fabrication. *J. Micro/Nanolithography MEMS MOEMS* **2013**, *12*, 041203. [CrossRef]
24. Heo, J.; Min, H.; Lee, M. Laser micromachining of permalloy for fine metal mask. *Int. J. Precis. Eng. Manuf. —Green Technol.* **2015**, *2*, 225–230. [CrossRef]
25. Lee, J.H.; Yoon, J.W.; Kim, I.G.; Oh, J.S.; Nam, H.J.; Jung, D.Y. Fabrication of polydimethylsiloxane shadow masks for chemical solution deposition of CdS thin-film transistors. *Thin Solid Films* **2008**, *516*, 6492–6498. [CrossRef]
26. Chung, I.Y.; Kim, J.D.; Kang, K.H. Ablation drilling of invar alloy using ultrashort pulsed laser. *Int. J. Precis. Eng. Manuf.* **2009**, *10*, 11–16. [CrossRef]
27. Cheng, J.Y.; Wei, C.W.; Hsu, K.H.; Young, T.H. Direct-write laser micromachining and universal surface modification of PMMA for device development. *Sens. Actuators B Chem.* **2004**, *99*, 186–196. [CrossRef]
28. Nayak, N.C.; Lam, Y.C.; Yue, C.Y.; Sinha, A.T. CO_2-laser micromachining of PMMA: The effect of polymer molecular weight. *J. Micromech. Microeng.* **2008**, *18*, 095020. [CrossRef]
29. Yuan, D.; Das, S. Experimental and theoretical analysis of direct-write laser micromachining of polymethyl methacrylate by CO_2 laser ablation. *J. Appl. Phys.* **2007**, *101*, 024901. [CrossRef]
30. Knowles, M.R.H.; Rutterford, G.; Karnakis, D.; Ferguson, A. Micro-machining of metals, ceramics and polymers using nanosecond lasers. *Int. J. Adv. Manuf. Technol.* **2007**, *33*, 95–102. [CrossRef]
31. Li, J.; Ananthasuresh, G.K. A quality study on the excimer laser micromachining of electro-thermal-compliant micro devices. *J. Micromech. Microeng.* **2000**, *11*, 38–47. [CrossRef]
32. Ihlemann, J.; Rubahn, K. Excimer laser micro machining: Fabrication and applications of dielectric masks. *Appl. Surf. Sci.* **2000**, *154*, 587–592. [CrossRef]

33. Pfleging, W.; Bernauer, W.; Hanemann, T.; Torge, M. Rapid fabrication of microcomponents—UV-laser assisted prototyping, laser micro-machining of mold inserts and replication via photomolding. *Microsyst. Technol.* **2002**, *9*, 67–74. [CrossRef]
34. Shiu, P.P.; Knopf, G.K.; Ostojic, M.; Nikumb, S. Rapid fabrication of tooling for microfluidic devices via laser micromachining and hot embossing. *J. Micromech. Microeng.* **2008**, *18*, 025012. [CrossRef]
35. Ricciardi, G.; Cantello, M.; Mariotti, F.; Castelli, P.; Giacosa, P. Micromachining with Excimer Laser. *CIRP Ann.* **1998**, *47*, 145–148. [CrossRef]
36. Teixidor, D.; Thepsonthi, T.; Ciurana, J.; Özel, T. Nanosecond pulsed laser micromachining of PMMA-based microfluidic channels. *J. Manuf. Process.* **2012**, *14*, 435–442. [CrossRef]
37. Liu, X.; Du, D.; Mourou, G. Laser ablation and micromachining with ultrashort laser pulses. *IEEE J. Quantum Electron.* **1997**, *33*, 1706–1716. [CrossRef]
38. Oblov, K.; Ivanova, A.; Soloviev, S.; Samotaev, N.; Lipilin, A.; Vasiliev, A.; Sokolov, A. Fabrication of Microhotplates Based on Laser Micromachining of Zirconium Oxide. *Phys. Procedia* **2015**, *72*, 485–489. [CrossRef]
39. Moser, R.; Kunzer, M.; Goßler, C.; Köhler, K.; Pletschen, W.; Schwarz, U.T.; Wagner, J.H. Laser processing of gallium nitride-based light-emitting diodes with ultraviolet picosecond laser pulses. *Opt. Eng.* **2012**, *51*, 114301. [CrossRef]
40. Yeh, J.T.C. Laser ablation of polymers. *J. Vac. Sci. Technol. A* **1986**, *4*, 653. [CrossRef]
41. Kancharla, V.V.; Chen, S. Fabrication of Biodegradable Polymeric Micro-devices using Laser Micromachining. *Biomed. Microdevices* **2002**, *4*, 105–109. [CrossRef]
42. Lorazo, P.; Lewis, L.J.; Meunier, M. Thermodynamic pathways to melting, ablation, and solidification in absorbing solids under pulsed laser irradiation. *Phys. Rev. B* **2006**, *73*, 134108. [CrossRef]
43. Luk'Yanchuk, B.; Bityurin, N.; Himmelbauer, M.; Arnold, N. UV-laser ablation of polyimide: From long to ultra-short laser pulses. *Nucl. Instrum. Methods Phys. Res. B* **1997**, *122*, 347–355. [CrossRef]
44. Zhang, Y.; Zhang, D.; Wu, J.; He, Z.; Deng, X. A thermal model for nanosecond pulsed laser ablation of aluminum. *AIP Adv.* **2017**, *7*, 075010. [CrossRef]
45. DeHoff, R. *Thermodynamics in Materials Science*, 2nd ed.; CRC Press: Boca Raton, FL, USA, 2006.
46. Taylor, R.E. *CINDAS Data Series on Materials Properties: Thermal Expansion of Solid*; ASM International: Materials Park, OH, USA, 1998; Volume 1–4.
47. Wang, Y.; Shen, N.; Befekadu, G.K.; Pasiliao, C.L. Modeling pulsed laser ablation of aluminum with finite element analysis considering material moving front. *Int. J. Heat Mass Transf.* **2017**, *113*, 1246–1253. [CrossRef]
48. Cheng, J.; Liu, C.S.; Shang, S.; Liu, D.; Perrie, W.; Dearden, G.; Watkins, K. A review of ultrafast laser materials micromachining. *Opt. Laser Technol.* **2013**, *46*, 88–102. [CrossRef]
49. Huang, H.; Guo, Z. Ultra-short pulsed laser PDMS thin-layer separation and micro-fabrication. *J. Micromech. Microeng.* **2009**, *19*, 055007. [CrossRef]
50. Wang, H.; Wu, X.; Dong, P.; Wang, C.; Wang, J.; Liu, Y.; Chen, J. Electrochemical Biosensor Based on Interdigitated Electrodes for Determination of Thyroid Stimulating Hormone. *Int. J. Electroch. Sci.* **2014**, *9*, 12–21.
51. Hart, C.; Kundu, A.; Kumar, K.; Varma, S.J.; Thomas, J.; Rajaraman, S. Rapid nanofabrication of nanostructured interdigitated electrodes (nIDEs) for long-term in vitro analysis of human induced pluripotent stem cell differentiated cardiomyocytes. *Biosensors* **2018**, *8*, 88. [CrossRef]
52. Alexander, F.; Price, D.T.; Bhansali, S. Optimization of interdigitated electrode (IDE) arrays for impedance based evaluation of Hs 578T cancer cells. *J. Phys. Conf. Ser.* **2010**, *224*, 012134. [CrossRef]

 *micromachines*

 MDPI

Article

A Low Contact Impedance Medical Flexible Electrode Based on a Pyramid Array Micro-Structure

Song Wang [1], Jin Yan [2], Canlin Zhu [1], Jialin Yao [2], Qiusheng Liu [1] and Xing Yang [1],*

[1] The State Key Laboratory of Precision Measurement Technology and Instruments, Department of Precision Instrument, Tsinghua University, Beijing 100084, China; wangs18@mails.tsinghua.edu.cn (S.W.); 13141317443@163.com (C.Z.); liuqiusheng@tsinghua.edu.cn (Q.L.)
[2] School of Materials Science and Engineering, University of Science and Technology Beijing, Beijing 100083, China; 41703112@xs.ustb.edu.cn (J.Y.); yaojialin92@gmail.com (J.Y.)
* Correspondence: yangxing@tsinghua.edu.cn; Tel.: +86-10-62779064

Received: 30 October 2019; Accepted: 27 December 2019; Published: 1 January 2020

Abstract: Flexible electrodes are extensively used to detect signals in electrocardiography, electroencephalography, electro-ophthalmography, and electromyography, among others. These electrodes can also be used in wearable and implantable medical systems. The collected signals directly affect doctors' diagnoses of patient etiology and are closely associated with patients' life safety. Electrodes with low contact impedance can acquire good quality signals. Herein, we established a method of arraying pyramidal microstructures on polydimethylsiloxane (PDMS) substrates to increase the contact area of electrodes, and a parylene transitional layer is coated between PDMS substrates and metal membranes to enhance the bonding force, finally reducing the impedance of flexible electrodes. Experimental results demonstrated that the proposed methods were effective. The contact area of the fabricated electrode increased by 18.15% per unit area, and the contact impedance at 20 Hz to 1 kHz scanning frequency ranged from 23 to 8 kΩ, which was always smaller than that of a commercial electrode. Overall, these results indicated the excellent performance of the fabricated electrode given its low contact impedance and good biocompatibility. This study can also serve as a reference for further electrode research and application in wearable and implantable medical systems.

Keywords: flexible electrode; polydimethylsiloxane (PDMS); pyramid array micro-structures; low contact impedance

1. Introduction

Electrodes are important components of implantable or wearable medical monitoring systems. They are used to collect electrocardiography (ECG) and electroencephalogram (EEG) signals. If the collected signals are inaccurate, patient treatment may be delayed, their conditions worsened, or their lives endangered. Therefore, high-quality electrodes are significant for precisely monitoring the physiological parameters of the human body. At present, ECG, EEG, electromyogram (EMG), or electro-ophthalmogram (EOG) signals are measured on a standard silver/silver chloride (Ag/AgCl) wet electrode for medical diagnosis [1]. Such electrodes usually reduce the contact impedance between the skin and electrode by using conductive gels [1]. Standard Ag/AgCl wet electrodes enable low electrode–skin contact impedance and the acquisition of signals with high stability and repeatability. However, the conductive gel may cause problems such as becoming dry with time, which increases the contact impedance and aggravates measurement errors [2]. Conductive gels can also sometimes cause skin irritations or allergic rashes [3,4].

The above problems have prompted researchers to explore dry electrodes with other conductive materials [4,5] and the use of microelectromechanical system technology [6]. For example, Fei Yu et al. (2012) prepared a flexible dry electrode based on parylene substrate to collect ECG signals in zebrafish; their electrode exhibits a good signal-to-noise ratio [7]. Electrode impedance is a criterion for evaluating electrode performance. A smaller impedance corresponds to better collected-signal quality [8]. Some progress has been made in the study of dry electrodes, but some challenges remain, such as those related to reducing the impedance of flexible dry electrodes. To achieve such reduction and obtain high-quality signals, this study theoretically analyzed the impedance of electrodes and the electrode–skin contact impedance. We then proposed a method of reducing electrode impedance by fabricating pyramidal array microstructures on polydimethylsiloxane (PDMS) substrate and adding a parylene transition layer between PDMS and metal film. Finally, the impedance of the flexible electrode was experimentally tested and analyzed.

2. Theoretical Analysis of Flexible Electrode Based on Micro-structure

2.1. Theoretical Analysis of Selecting Material of Flexible Electrode Substrate

To reduce electrode impedance and enhance the bonding force between substrate and metal film, we selected substrate materials for flexible electrodes on the basis of two aspects: one is the bonding degree between substrate and biological tissues, and the other is flexibility that should be similar to that of biological tissues. The first basis for substrate selection considers the degree of adherence between the substrate and biological tissues. A material should have flexibility similar to that of biological tissues, i.e., they should be easily bent and stretched. The second basis for substrate selection considers the flexibility, i.e., polymer materials with good elasticity should be selected. Young's modulus of PDMS in polymer (0.4–1.0 MPa) is much smaller than that of other polymer materials, such as parylene (2400–3200 MPa). Thus, PDMS is easier to bend and deform and is more suitable for the fabrication and bonding of flexible substrates with large thickness. At the same time, PDMS is inexpensive and simple to manufacture; it can be prepared by simple spin coating. These characteristics indicate the suitability of PDMS materials as the substrate material for flexible electrodes. However, considering the impedance stability and thermal expansion between substrate and metal, two major problems arise in selecting PDMS material as substrate to directly bond with metal. On one hand, a porous PDMS material has strong permeability, and an electrode easily absorbs water owing to capillary action, which can lead to changes in electrode impedance. On the other hand, the weak adhesion forces between PDMS and metal confer difficulty in the deposition and transfer of a metal electrode onto PDMS [9]. Considering that the thermal-expansion coefficient of PDMS ($\alpha_{PDMS} \approx 20 \times 10^{-5}$ K) relatively differs from that of gold ($\alpha_{Au} \approx 1.42 \times 10^{-5}$ K) [10], the metal film on PDMS substrate during electrode fabrication and storage is prone to cracking because of changes in temperature. Consequently, electrode impedance increases.

To address the two problems, a layer of parylene can be introduced between PDMS substrate and gold electrode. On one hand, the water permeability of parylene is very low, so water cannot easily penetrate parylene. The combination of parylene and PDMS can offset the shortcomings of the water absorption of PDMS and overcome the change in electrode impedance. On the other hand, parylene is also a common material for preparing flexible substrates. Its thermal-expansion coefficient ($\alpha_{parylene} \approx 3.5 \times 10^{-5}$ K) is similar to that of gold [10], and it is hardly affected by cracks caused by temperature factors after bonding with gold. The resulting PDMS–parylene layer possesses a more favorable combination of stiffness and flexibility that can improve film manageability, and the fabrication process is straightforward [11]. This process overcomes the disadvantages of the commonly used PDMS flexible electrode, which induces impedance changes owing to its easy water absorption and crack and wrinkle formation.

2.2. Theoretical Analysis of Contact Impedance of Flexible Electrode–skin

To analyze electrode–skin contact impedance, selecting a suitable impedance circuit model is very important. We utilize the H. Feriberger equivalent circuit model, which is extensively used in medicine [12], as is shown in the following Figure 1:

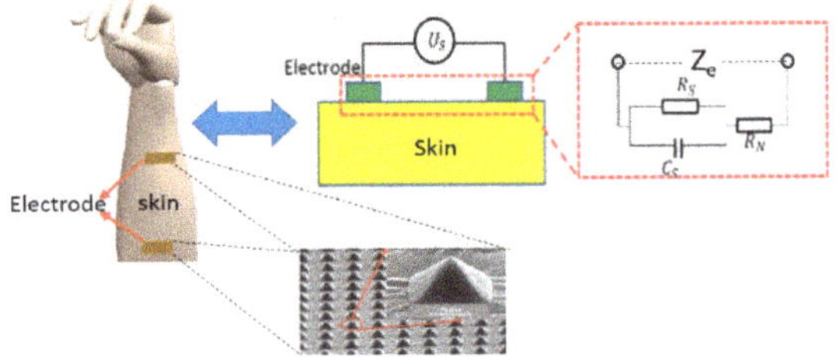

Figure 1. Skin–electrode contact impedance equivalent circuit diagram.

Where R_c, R_s, R_N, and Z_e are the impedance of capacitance, impedance of electrode - skin, internal fixed impedance of human body (a constant value) [12] and the total impedance(a series–parallel structure of skin impedance, skin capacitance, and internal impedance of human body). u_s is the given constant voltage, C_s is the skin capacitance between two electrodes. ε_0 and ε_r are the vacuum dielectric constant and skin dielectric constant, respectively. A is the contact area of flexible electrodes with the skin, d is the distance between flexible electrodes, ρ is the skin impedance coefficient, ω is the angular frequency. The related formulas are as follows:

$$C_s = \frac{\varepsilon_0 \times \varepsilon_r \times A}{d} \tag{1}$$

$$R_s = \rho \times \frac{d}{A} \tag{2}$$

$$R_c = \frac{1}{j\omega \times C_s} = \frac{d}{j\omega \times \varepsilon_0 \times \varepsilon_r \times A} \tag{3}$$

$$R_c//R_s = \frac{\frac{\rho d}{A} \times \frac{d}{j\omega\varepsilon_0\varepsilon_r A}}{\frac{\rho d}{A} + \frac{d}{j\omega\varepsilon_0\varepsilon_r A}} = \frac{\rho d}{A(1 + j\omega\varepsilon_0\varepsilon_r\rho)} \tag{4}$$

$$Z_e = R_c//R_s + R_N = \frac{R_s}{1 + j\omega R_s C_s} + R_N = \frac{\rho d}{A(1 + j\omega\varepsilon_0\varepsilon_r\rho)} + R_N \tag{5}$$

As previously mentioned and analyzed, a smaller electrode–skin contact impedance corresponds to a higher quality of collected signals. Thus, the contact impedance directly affects electrode performance. According to Equations (2) and (3), a larger contact area of electrodes with the skin means a smaller impedance of the electrode–skin contact capacitance. Moreover, the total electrode–skin contact impedance decreases according to Equations (4) and (5), so increasing the contact area between the electrode and skin can produce high-quality signals. To increase the contact area with skin per unit area, some special microstructures were designed and fabricated on the electrode contact surface. These microstructures were micro-pyramidal arrays with a base of 8 μm × 8 μm and an 8 μm interval between each pyramidal microstructure. This dense pyramidal array effectively increased the contact area with skin per unit area without increasing the electrode size. The images of the pyramidal microstructures are shown in Figure 2.

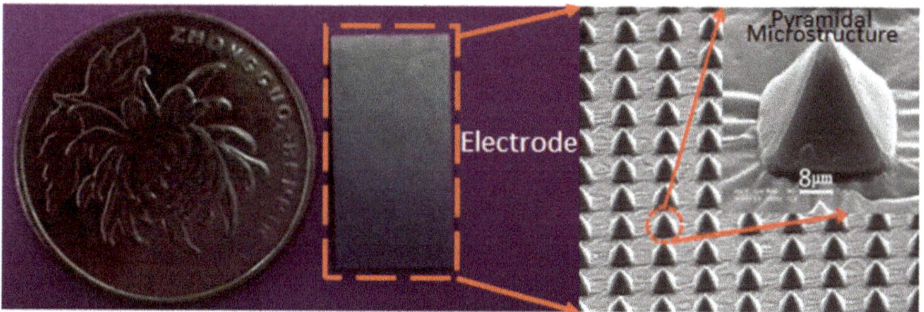

Figure 2. The SEM photographs of pyramidal microstructure arrays.

In this study, 781,250 microstructural pyramids were fabricated on the surface of PDMS flexible dry electrode with dimensions of 2 cm (L) × 1 cm (W). The total surface area of the flexible dry electrode with micro-pyramidal array was 2,363,072 μm^2, which was 363,072 μm^2 larger than that of the planar flexible dry electrode. Therefore, the contact area of the flexible dry electrode can be increased by 18.15% through the fabrication of pyramidal microstructures on PDMS substrates, thereby effectively reducing the contact impedance between electrode and skin.

3. Fabrication of Flexible Electrodes

The fabrication of flexible electrodes based on PDMS pyramidal arrays was divided into many steps. Firstly, a mold of pyramidal array pits was fabricated by etching on silicon wafers. The etching method was as follows. Apply reactive ion etching (RIE) technology to etch silicon nitride with a thickness of 300 nm on the upper surface of silicon wafer for 30 min. Then use the anisotropic wet etching method of silicon: the silicon wafer was put into 33.3% potassium hydroxide solution for etching, while heating in water bath at 80 °C for 10 min.

Secondly, fabricate PDMS flexible substrate with pyramid microstructure. The method was as follows. Put the silicon template with pyramid microstructure into acetone, ethanol, deionized water and mold release agent in turn and perform ultrasonic treatment for 10 min respectively. Then take out the silicon wafer and dry it to obtain the pretreated silicon template. At the same time, we mixed the PDMS liquid elastomer and curing agent uniformly with the mass ratio of 10:1 and placed the mixture in a vacuum chamber to be pumped for 15 min so that bubbles in the mixture could be fully discharged. After that, pour the PDMS mixture on the pretreated silicon template with pyramid microstructure. At the same time, spin-coat the PDMS mixture uniformly with a spin coater. Next, place the silicon wafer with PDMS mixture in an oven for baking at 70 degree for 2 h to solidify the PDMS mixture.

After solidification, strip the solidified PDMS from the silicon wafer to obtain a 310 μm thick PDMS flexible substrate with pyramid microstructure, and then a 3.5-μm-thick parylene-C layer was fabricated on the prepared PDMS substrates by chemical vapor deposition. Finally, a metal film was magnetron sputtered onto parylene-C. In this study, gold was selected as the metal layer material with 40 nm thickness. This fabrication was simple and inexpensive as the fabrication of PDMS film substrate did not require any special deposition equipment. The fabricated PDMS film was also very flexible. The detailed fabrication process of flexible electrodes is shown in Figure 3.

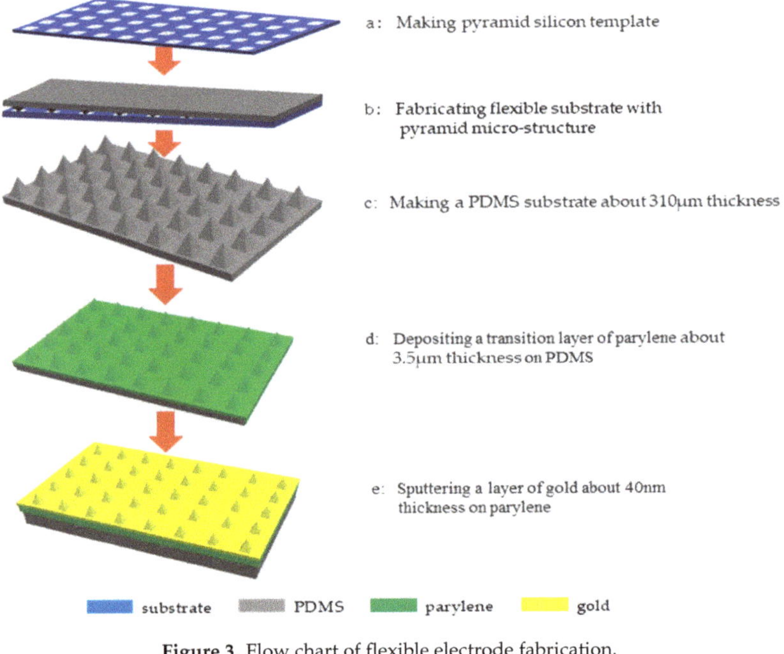

Figure 3. Flow chart of flexible electrode fabrication.

4. Testing Performance of Electrode

4.1. Electrode Self-Impedance Testing

In this experiment, two types of PDMS flexible electrodes with and without a parylene transition layer were compared and analyzed. The impedance of the electrode was tested at 20 Hz. The impedance of PDMS pyramidal array electrode without the parylene transition layer was 74.94 MΩ, and that of PDMS pyramidal array electrode with the parylene transition layer was 49.52 Ω. After measuring the electrode impedance, we also observed the morphology of the two electrodes under a microscope, and the results were as follows.

According to Figure 4, several cracks were present on the metal film of the electrode without the parylene transition layer, and some of the micro-pyramid structures collapsed and cracked. Meanwhile, the contact surface morphology of the PDMS electrode with the parylene transition layer was intact, barely producing micro-cracks. This finding was due to the thermal-expansion coefficient of PDMS greatly differing from that of gold, which caused the destruction of gold on the pyramid structure when the metal was directly sputtered. The thermal-expansion coefficient of parylene is close to that of gold, but pyramid deformation can hardly be transmitted and did not affect the gold film during magnetron sputtering after the parylene transition layer was added. The experiments demonstrated that adding the parylene transition layer between PDMS and gold can effectively prevent the breakage of the metal film, thereby ensuring the small and stable impedance of PDMS electrode with pyramidal array microstructure.

4.2. Characterization and Testing Performance of Electrode–Skin Contact Impedance

To verify whether the fabricated flexible electrode based on PDMS pyramidal array microstructure had low electrode–skin contact impedance, we used a commercial electrode (Tianrun Sunshine (Beijing) Commercial Products Co., Ltd. Beijing, China) with the same size to perform comparative experiments. For impedance analysis, a Precision Impedance Analyzer (Wayne Kerr 6500B, made in UK,

Wayne Kerr Electronics, West Sussex, UK) was used. We obtained the spectrum of the electrode–skin contact impedance for flexible and commercial electrodes separately by scanning the frequency, whose interval was set from 20 Hz to 1 kHz. The driving voltage was set at 1 V. The two electrodes were attached to the left arm of a test person, and their center distance was kept at 10 cm. The experimental data were analyzed using MATLAB (R2017b, The MathWorks, Inc., Natick, MA, USA).

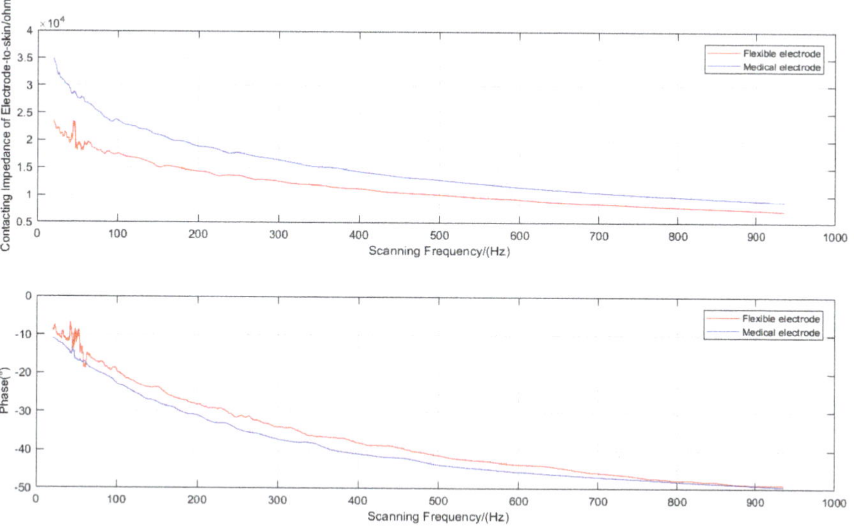

(a) **(b)**

Figure 4. Scanning electron microscope (SEM) photographs of metal films on electrodes (**a**) Electrodes without parylene transition layer (**b**) Electrodes containing the parylene transition layer.

Figure 5 shows that the difference between the fabricated flexible electrode and commercial electrode was obvious at low frequencies; both their contact impedance and phase gradually decreased with increased scanning frequency. We found that although the scanning frequency was affected by power-frequency interference at around 50 Hz, the impedance value of the fabricated flexible electrode at any frequency was always smaller than that of the commercial electrode, indicating its superiority over the commercial electrode.

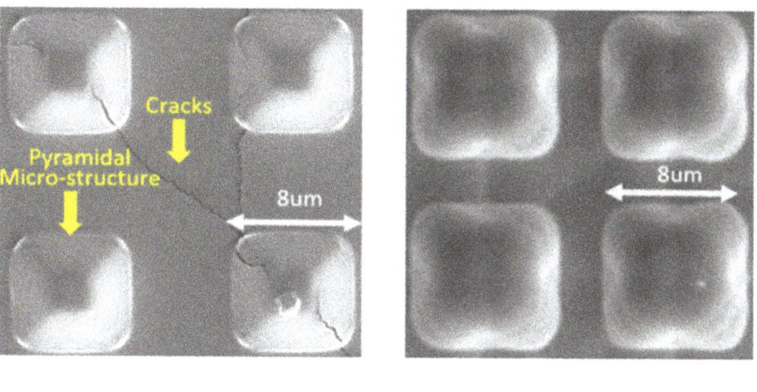

Figure 5. Comparison of skin contact impedance between the commercial electrodes and the flexible electrodes with pyramidal array micro-structure.

5. Conclusions

This study proposed a simple and efficient method of fabricating a low-contact-impedance and flexible electrode with pyramid array microstructure. By fabricating the micropyramid array on the electrode surface, the contact area of the fabricated electrode with the skin increased, which indirectly reduced the impedance of flexible electrodes. It is shown that the contact area of fabricated electrodes is increased by 18.15% per unit area with adding pyramid array micro-structure on the surface of the electrode through calculations.

Upon analysis of the contact impedance of the flexible electrode–skin circuit equivalent model, we studied the simple process of fabricating the flexible electrode with pyramid array microstructure. Experimental results demonstrated that the contact impedance ranged from 23 to 8 kΩ at 20 Hz to 1 kHz scanning frequency, which was always smaller than that of the commercial electrode. Therefore, the proposed method is effective and the fabricated flexible dry electrode has good performance in terms of low impedance and good biocompatibility, indicating the potential applications of the electrode in EOG, ECG, and EMG, among others.

Author Contributions: X.Y. and Q.L. conceived and designed the experiments and provided experimental instruments. S.W. and C.Z. performed the experiments. S.W. analyzed the data. X.Y., S.W., J.Y. (Jin Yan) and C.Z. wrote the paper. J.Y. (Jialin Yao) took part in some experiments. All authors have read and agreed to the published version of the manuscript.

Funding: This work was financially supported by the Beijing Natural Science Foundation (Grant No. L172040) and National Natural Science Foundation of China Project (Grant No. 61671271, 51705283 and 51735007).

Acknowledgments: I would like to acknowledge Professor Yonggui Dong in Department of Precision Instruments, Tsinghua University, for his detailed guidance and many valuable opinions on scientific research of this paper.

Conflicts of Interest: The authors declare no conflict of interest. The funders had no role in the design of the study; in the collection, analyses, or interpretation of data; in the writing of the manuscript, or in the decision to publish the results.

References

1. Chlaihawi, A.A.; Narakathu, B.B.; Emamian, S.; Bazuin, B.J.; Atashbar, M.Z. Development of printed and flexible dry ECG electrodes. *Sens. Bio-Sens. Res.* **2018**, *20*, 9–15. [CrossRef]
2. Xu, S.; Dai, M.; Xu, C.; Chen, C.; Tang, M.; Shi, X.; Dong, X. Performance evaluation of five types of Ag/AgCl bio-electrodes for cerebral electrical impedance tomography. *Ann. Biomed. Eng.* **2011**, *39*, 2059–2067. [CrossRef] [PubMed]
3. Yokus, M.A.; Jur, J.S. Fabric-based wearable dry electrodes for bodycontact surface bio-potential recording. *IEEE Trans. Biomed. Eng.* **2015**, *63*, 423–430. [CrossRef] [PubMed]
4. Myers, A.; Huang, H.; Zhu, Y. Wearable silver nanowire dry electrodes for electro-physiological sensing. *Roy. Soc. Chem.* **2015**, *5*, 11627–11632.
5. Zhou, Y.; Ding, X.; Zhang, J.; Duan, Y.; Hu, J.; Yang, X. Fabrication of conductive fabric as textile electrode for ECG monitoring. *Fibers Polym.* **2014**, *15*, 260–2264. [CrossRef]
6. Jung, H.; Moon, J.; Baek, D.; Lee, J.; Choi, Y.; Hong, J.; Lee, S. CNT/PDMS composite flexible dry electrodes for long-term ECG monitoring. *IEEE Trans. Biomed. Eng.* **2012**, *59*, 1472–1479. [CrossRef] [PubMed]
7. Yu, F.; Zhao, Y.; Gu, J.; Quigley, K.L.; Chi, N.C.; Tai, Y.C.; Hsiai, T.K. Flexible microelectrode arrays to interface epicardial electrical signals with intracardial calcium transients in zebrafish hearts. *Biomed. Microdevices* **2012**, *14*, 357–366. [CrossRef] [PubMed]
8. Ying, M. Research on Key Technologies of Intelligent ECG Monitoring System Based on Flexible MEMS Dry Electrode Array. Ph.D. Thesis, Shanghai Jiao Tong University, Shanghai, China, 2015.
9. Koh, D.; Wang, A.; Schneider, P.; Bosinski, B.; Oh, K. Introduction of a Chemical-Free Metal PDMS Thermal Bonding for Fabrication of FlexibleElectrode by Metal Transfer onto PDMS. *Micromachines* **2017**, *8*, 280.

10. Chou, N.; Yoo, S.; Kim, S. A Largely Deformablecontact surface Type Neural Electrode Array Based on PDMS. *IEEE Trans. Neural Syst. Rehabil. Eng.* **2013**, *21*, 544–553. [CrossRef] [PubMed]
11. Ochoa, M.; Wei, P.; Wolley, A.J.; Otto, K.J.; Ziaie, B. A Hybrid PDMS-Parylene Subdural Multi-Electrode Array. *Biomed. Microdevices* **2013**, *15*, 437–443. [CrossRef] [PubMed]
12. Yuan, G. Research and Application of Human Body Effect of Low Frequency Current. Master's Thesis, Central China Normal University, Wuhan, China, 2006.

Article

Rapid and Sensitive Detection of Bisphenol A Based on Self-Assembly

Haoyue Luo, Xiaogang Lin *, Zhijia Peng, Min Song and Lifeng Jin

Key Laboratory of Optoelectronic Technology and Systems of Ministry of Education of China,
Chongqing University, Chongqing 400044, China; 20133029@cqu.edu.cn (H.L.); 201808021026@cqu.edu.cn (Z.P.);
201908131078@cqu.edu.cn (M.S.); 20152394@cqu.edu.cn (L.J.)
* Correspondence: xglin@cqu.edu.cn

Received: 27 November 2019; Accepted: 27 December 2019; Published: 30 December 2019

Abstract: Bisphenol A (BPA) is an endocrine disruptor that may lead to reproductive disorder, heart disease, and diabetes. Infants and young children are likely to be vulnerable to the effects of BPA. At present, the detection methods of BPA are complicated to operate and require expensive instruments. Therefore, it is quite vital to develop a simple, rapid, and highly sensitive method to detect BPA in different samples. In this study, we have designed a rapid and highly sensitive biosensor based on an effective self-assembled monolayer (SAM) and alternating current (AC) electrokinetics capacitive sensing method, which successfully detected BPA at nanomolar levels with only one minute. The developed biosensor demonstrates a detection of BPA ranging from 0.028 µg/mL to 280 µg/mL with a limit of detection (LOD) down to 0.028 µg/mL in the samples. The developed biosensor exhibited great potential as a portable BPA biosensor, and further development of this biosensor may also be useful in the detection of other small biochemical molecules.

Keywords: alternating current (AC) electrokinetics; bisphenol A; self-assembly; biosensor

1. Introduction

Bisphenol A (BPA) is an important organic chemical raw material, which is widely used in the production of plastic products and fire retardant. BPA is an endocrine disruptor, which can mimic human hormones and maybe lead to negative health effects [1,2]. Studies have shown that BPA can contact humans through the skin, respiratory tract, digestive tract, and other channels. After BPA enters the body, it combines with intracellular estrogen receptors and produces estrogenic or anti-estrogenic effects through a variety of reaction mechanisms, thereby affecting endocrine, reproductive, and nervous systems, as well as causing cancer and other adverse effects [3–5]. Therefore, a rapid and sensitive detection method of BPA is of great significance.

At present, the detection methods of BPA mainly include liquid chromatography with an electrochemical method [6], chromatography-mass spectrometry [7,8], surface-enhanced Raman scattering [9], enzyme-linked immunosorbent assay (ELISA) [10], etc. These methods are complicated and require expensive equipment with professional personnel. The movement of biomolecular molecules in the detection process relies on natural diffusion, which is time-consuming and cannot meet the requirements for the rapid detection of BPA. For example, Pasquale et al. [11] have proposed a method to determine BPA levels in fruit juices by liquid chromatography coupled to tandem mass spectrometry. However, it cost about 15 min to accomplish the detection of BPA. Sheng et al. [12] have developed an optical biosensor based on fluorescence, but they required the addition of labels to generate the sensor response. Xue et al. [13] have reported a novel SPR biosensor that combines a binding inhibition assay with functionalized gold nanoparticles to allow for the detection of trace concentrations of BPA. However, the biosensor detected BPA in 60 min, which was too long for rapid detection. Recently, various types of electrochemical sensors have attracted more attention. These electrochemical sensors are always modified via different sensing materials, such as molecularly imprinted polymers [14], carbon nanotube [15,16], graphene [17,18], nanocomposites [19,20], and metal composites [21]. For example, Maryam et al. [22] have used an electroactive label-based aptamer to detect bisphenol A in serum samples with the linear range of 0.2–2 nM. Lee et al. [23] have reported a simple and label-free colorimetric aptasensor for BPA detection. They have used the spectroscopic methods, which was time-consuming and could not achieve point-of-care testing. Inroga et al. [24] have used gold leaf-like microstructures to develop a tyrosinase-based biosensor for bisphenol A detection with the limit of detect of 0.077 μmol/L. Liu et al. [25] have developed an electrochemical enzyme biosensor bearing biochar nanoparticles as signal enhancers for bisphenol A detection in water. Peng et al. [26] have developed a signal-enhanced lateral flow strip biosensor for the visual and quantitative detection of BPA based on the poly amidoamine (PAMAM) binding with antibody. The detection would cost 10 min. Liu et al. [27] have developed a solution-gated graphene transistor (SGGT) modified with DNA molecules in a microfluidic system for a recycling detection of BPA. Compared with most of the recent publications, which have reported the electrochemical method to achieve the detection of BPA, the results of this work may not be better than them. However, in this work, we have demonstrated a novel method to detect BPA based on self-assembly technology and alternating current (AC) electrokinetics effect. To our knowledge, there has not been any report that detects BPA with self-assembly technology and an AC electrokinetics effect. So, it has the tremendous potential to detect BPA more sensitively with a lower limit of detection in the near future.

In this work, a rapid, highly sensitive BPA biosensor based on self-assembly technology and AC electrokinetics (ACEK) method was developed, which could accomplish the rapid and sensitive detection of BPA. The surface of the interdigital electrode was functionalized with the antibody of BPA via self-assembly and used for the specific capture of antigen of BPA. ACEK is used to realize the enrichment of biomolecules through the manipulation of microfluidic and nanoparticles by an AC electric field [28,29]. Compared with traditional detection methods, the ACEK method has the advantages of less sample consumption, faster detection, and a simpler procedure with an appropriate AC signal applied to microelectrode sensors in sample fluids. Then, ACEK microflows accelerate molecules of BPA binding [30,31]. With the enrichment of the target molecules, the interfacial capacitance of the interdigital electrode will change. Therefore, the change of the interfacial capacitance can be used to characterize the concentration of sample solution. Compared with the previous work [32], there are some differences in this work. Firstly, in this work, we used the BPA antibody to achieve the detection of BPA, while an aptamer probe was used to detect BPA in the previous work. Secondly, we used the self-assembly technology to form a self-assembled monolayer with the mercaptan containing alkyl chains immobilized on the gold electrodes via Au–S bonds. Then, N-hydroxysuccinimide (NHS) and 1-(3-dimethylaminopropyl)-3-ethyl carbon diimide hydrochloride (EDC) were used as the crosslinking agent to assist in the formation of amide bonds between the carboxyl group of the self-assembled monolayer and the amino group of the BPA antibody. With these processes, the BPA antibody was immobilized onto the electrodes stably, while a BPA aptamer specific to BPA with the 5′ thiol modifier C6 SH (Fisher Scientific, PA) was directly immobilized onto the electrodes in the previous work. Thirdly, in this work, the interdigital microelectrodes had interdigitated arrays with widths of 10 μm separated by 10-μm gaps. However, the interdigital microelectrodes had interdigitated arrays with widths of 6 μm separated by 6-μm gaps in the previous work. Lastly, in this work, a 10 kHz AC signal was applied to the microelectrodes, but a 20 kHz AC signal was applied to the microelectrodes in the previous work.

With the improved biosensor design and use of an antigen and antibody, we successfully detected BPA at the nanomolar level within only one minute. Moreover, the developed BPA biosensor is expected to be used in the detection of biological fluids. It is of great significance to promote the detection of other small biochemical molecules.

2. Materials and Methods

2.1. Materials and Reagents

The buffer solution 10 × PBS was purchased from Solarbio (Beijing, China). The 10 × PBS was diluted in deionized water to make the working solution at 0.05 × PBS, 0.01 × PBS for diluting other substances. At the same time, the 0.01 × PBS was used as the background solution. 11-mercaptoundecanoic acid (MUA) was purchased from Yuanye Bio-Technology Co., Ltd., (Shanghai, China) The MUA was dissolved in anhydrous ethanol to prepare 5 mmol/L of MUA solution for forming the gold–sulfur bond. 1-(3-dimethylaminopropyl)-3-ethyl carbon diimide hydrochloride (EDC) was obained from Sigma (St. Louis, MI, USA); EDC was dissolved in 0.05 × PBS to make 0.4 mol/L EDC solution. N-hydroxysuccinimide (NHS) was acquired from biotopped (Beijing, China), and NHS was dissolved in 0.05 × PBS to prepare 0.1 mol/L NHS solution. Then, we mixed the two solution 1 to 4. Ethanolamine was purchased from MACKLIN (Shanghai Macklin Biochemical, Ltd., Shanghai, China). Ethanolamine was diluted with 0.05 × PBS to make 1 mol/L solution for closing. The BPA antibody and antigen were purchased from QF Biosciences Co., Ltd., (Shanghai, China). The antibody was diluted with 0.05 × PBS to make two concentrations of 5.3 µg/L and 5.3 µg/mL, and the antigen was diluted in 0.01 × PBS to make the solution that included 0.028 µg/mL, 0.28 µg/mL, 2.8 µg/mL, 28 µg/mL, and 280 µg/mL for testing. Antigen and antibody ware stored at −20 °C. In order to guarantee the accuracy of detection, preparation of the solution was carried out on a clean bench.

2.2. Preparation of Interdigital Microelectrodes

In this work, the interdigital microelectrodes were fabricated on silicon wafers, which had interdigitated arrays with widths of 10 µm separated by 10-µm gaps [32]. Before the detection, the interdigital microelectrodes should be modified with the following steps, as shown in Figure 1. Firstly, they are immersed in acetone for 4 min with ultrasonic cleaning, rinsed in absolute ethyl alcohol for 3 min with ultrasonic cleaning, rinsed in deionized water for 3 min with ultrasonic cleaning, and dried with a drying oven. Secondly, MUA was added to the surface of the gold electrodes for forming the Au–S bonds and realizing the self-assembled monolayer [33,34]. Then, sensors were placed in the incubator overnight with the temperature at 25 °C. Thirdly, before the EDC and NHS solution were added to the electrodes, the electrodes surface should be cleaned with absolute ethyl alcohol and blow dried with nitrogen; then, the sensors should be placed in an incubator for 2 h. After activation of the carboxyl group, the electrodes' surface should be cleaned with deionized water and blow dried with nitrogen; then, the chambers were pasted on the sensors. After that, 10 µL of antibody should be added to the surface of the sensors, which are then placed in an incubator for 3 h with the temperature at 37 °C. EDC and NHS were used as the crosslinking agent to assist in the formation of amide bonds between the carboxyl group of the self-assembled monolayer and the amino group of the BPA antibody. With these processes, the BPA antibody was immobilized onto the electrodes stably. Finally, in order to enhance the specificity of detection, ethanolamine was utilized to close the unbound active sites of the electrodes' surface. Then, the sensors were placed in an incubator for 1 h with the temperature at 25 °C. After the above steps, the modification of the electrodes surface was accomplished. The sensor needs to be cleaned after each sensing procedure; it is possible to wash off the antigen only and reuse the sensor, which indicates that the sensor has an expected reusability of five times before the performance of the sensor is believed to be dissatisfactory.

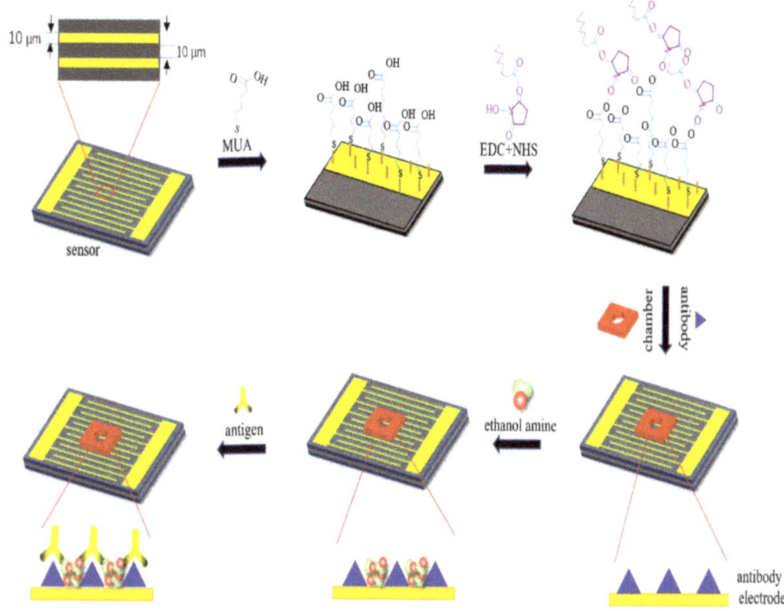

Figure 1. Representation of surface modification techniques on the interdigital electrodes' surface for the detection of bisphenol A (BPA).

2.3. Apparatus and Methods

An impedance analyzer of model IM3536 (HIOKI, Ueda, Japan) was a high-precision apparatus to detect the impedance, capacitance, and resistance of the modified interdigital electrode after being dropped different concentrations of antigens solution. Firstly, different frequencies (1 kHz, 10 kHz, 20 kHz) and different voltages (100 mV, 600 mV, 1.1 V) were applied to the experiments, respectively. Here, Figure 2 showed the relationship between the normalized capacitance and different concentrations of BPA with a 10-kHz AC signal of different voltages (100 mV, 600 mV, and 1.1 V) applied to the interdigital electrodes. Finally, a 10-kHz AC signal of 600 mV was selected to be applied to the interdigital electrodes via the impedance analyzer of model IM3536 as the measuring signal. At the same time, the voltage used in the experiments is rms. In this work, before detecting antigen concentrations, each sensor should detect the background solution. However, the measurements of the background solution are made in this work only and do not need to be used in commercial application. We detected the background solution as the blank control group to highlight the antibody–antigen binding response to the change in capacitance of electrodes. Then, 10 µL of different concentrations of antigens solution were added to interdigital electrodes. The impedance, capacitance, and resistance of the modified interdigital electrodes were detected within one minute. The normalized capacitance change rate of the sensor was computed to demonstrate antigen–antibody binding, which was shown with the slope of normalized capacitance versus time (%/min). Then, the slope was linearly fitted by the least square method. The normalized capacitance was computed as C_t/C_0, where C_t is the capacitance value at time t and C_0 is the capacitance value at time zero [32].

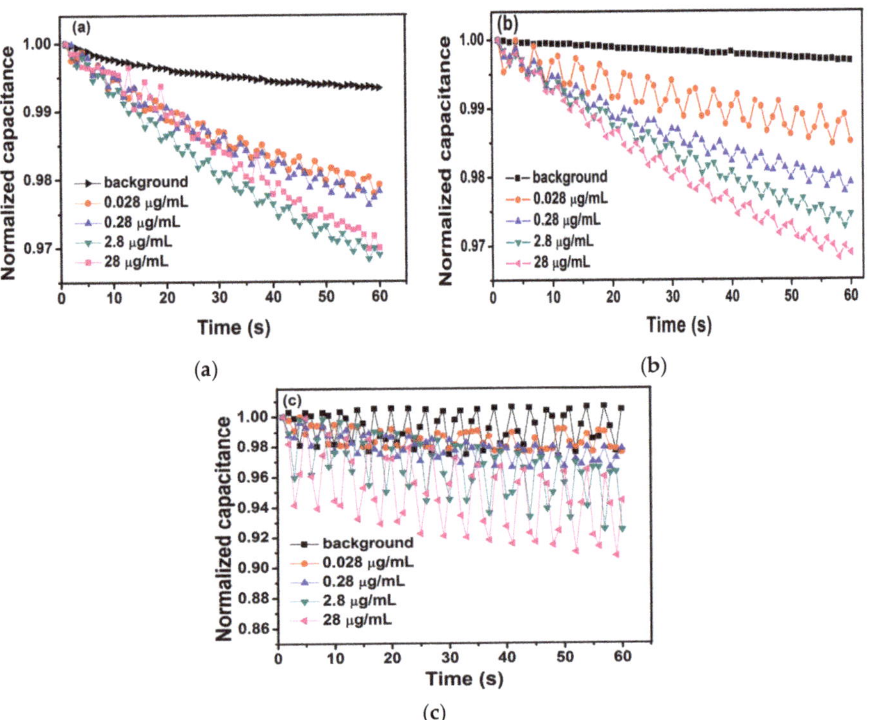

Figure 2. The relationship between the normalized capacitance and different concentrations of BPA with a 10-kHz AC signal of different voltages ((**a**) 100 mV, (**b**) 600 mV, and (**c**) 1.1 V) applied to the interdigital electrodes.

2.4. Electrical Double Layers

What happens when a solution comes into contact with a solid metal surface? Helmholtz built a model to attempt to explore this question [35]. He called it model H as the left half of the dotted line in Figure 3. As is shown in Figure 3, this model can be equivalent to a flat plate capacitor, and the relationship between the charge density (σ) on one side and the potential (V) difference between the two layers is described by the following equations [35,36], where d is the distance of center of the positive and negative charge.

$$\sigma = \frac{\varepsilon \varepsilon_0}{d} V \tag{1}$$

$$\frac{\partial \sigma}{\partial V} = C_H = \frac{\varepsilon \varepsilon_0}{d} \tag{2}$$

We can conclude the capacitance (C_H) of the flat plate capacitor from Equation (1). Hence, the H model can successfully describe the common electrochemical phenomenon with two basic equations. However, the Helmholtz layer shows an obvious defect, as shown in Figure 3. In the corollary of the equation, C_H is a constant, but in the experiment, there are spreading layers, leading to C_H not accurately describing the surface change. We describe the diffuse layer as C_D and call the electrical double layer (EDL) as C_d [36], which will be affected by the relative potential and electrolyte concentration. From this, we can deduce that C_d is equal to C_H in series with C_D. The value of C_d can be determined by the following equation.

$$C_d = \frac{C_H C_D}{C_H + C_D} \tag{3}$$

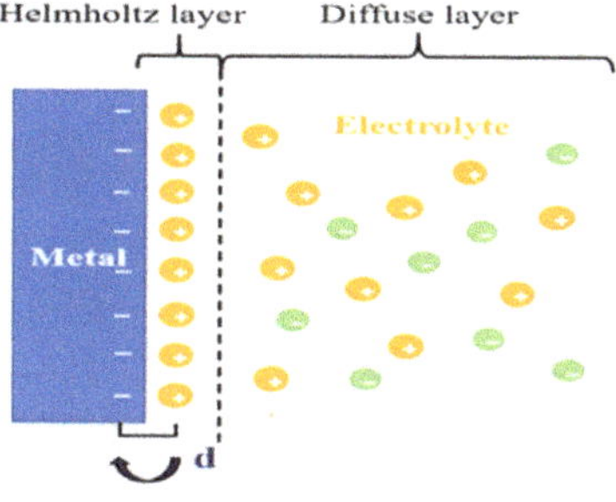

Figure 3. The double layers show the opposite charges equally distribute on both sides of the interface.

As mentioned above, both the charging and discharging processes of the electrical double layer (EDL) are similar to those of the parallel plate. When the electrolyte is added to the surface of interdigitated microelectrodes, the electrolyte will contact with the surface of the microelectrode in Figure 3. Thus, it can be equivalent to the parallel plate, as shown in Figure 4 [37]. When the electrode is bare, the interface capacitance of it can be described by the following equation.

$$C_{int,0} = \frac{A_{int}}{\dfrac{\lambda_d}{\varepsilon_s}} \tag{4}$$

where ε_s is the permittivity of the solution, A_{int} is the electrode area, and λ_d is the electrical double layer (EDL) thickness.

When antibodies are immobilized to the surface of microelectrodes via self-assembly, the interfacial capacitance C_{int} is expected to change to

$$C_{int,ab} = \frac{A_b}{\dfrac{\lambda_d}{\varepsilon_s} + \dfrac{d_{ab}}{\varepsilon_t}} \tag{5}$$

where ε_t is the permittivity of the antibody, A_b is the effective area after antibodies are immobilized to the surface of microelectrodes, and d_{ab} is the antibody thickness.

In the experiments, the solution including the antigens was added to the surface of interdigitated microelectrodes, on which the antibodies were immobilized. When the antigens bind to antibodies, the molecular deposition on the sensor surface become thicker, and the interfacial capacitance $C_{int,ab}$ will be expressed with

$$C_{int,ag} = \frac{A_g}{\dfrac{\lambda_d}{\varepsilon_s} + \dfrac{d_{ab}}{\varepsilon_t} + \dfrac{d_{ag}}{\varepsilon_p}} \tag{6}$$

where ε_p is the permittivity of the antigen, d_{ag} is the antigen thickness, and A_g is the effective area of the interfacial capacitor after the binding of an antigen to an antibody. Assume that the area of interfacial capacitance is equal before and after the binding. From Equation (6), we can see that the diminution of interfacial capacitance can be lead by the thickness increase of the dielectric layer. In this work, the biosensing utilizes the change of interfacial capacitance $C_{int,ab}$. Thus, the relative changes of interfacial capacitance are used to detect the specific binding of antigens to antibodies. The relative changes of interfacial capacitance are

$$\frac{\Delta C}{C_{int,ab}} = \frac{C_{int,ab} - C_{int,ag}}{C_{int,ab}} = -d_{ag}/(\frac{\varepsilon_p}{\varepsilon_s}\lambda_d + \frac{\varepsilon_p}{\varepsilon_t}d_{ab} + d_{ag}). \tag{7}$$

Consequently, the value of $\Delta C/C_{int,ab}$ can be used to detect the biomolecular interactions. Beyond that, measuring the value of $\Delta C/C_{int,ab}$ can overcome the experimental difference caused by the different surface roughness of each electrode and the difference of experimental treatment in each group, and improve the accuracy of experimental results.

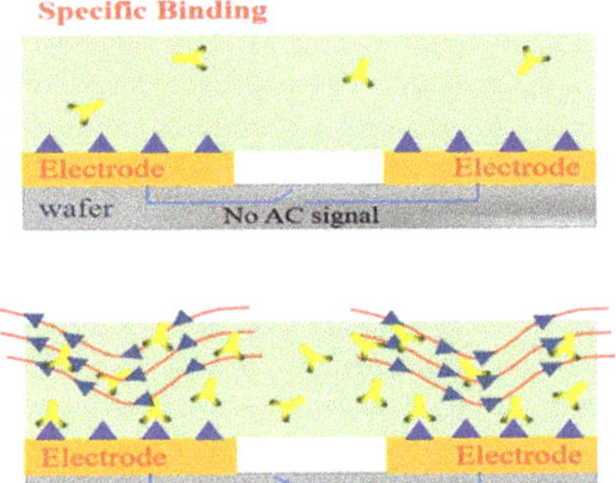

Figure 4. Changes on the electrode surface due to the specific binding of antigens to antibodies.

2.5. The Binding Mechanism of Interdigital Electrode Surface

In this work, we adopt the interdigital electrodes as the sensors. The interdigital microelectrodes are finger-shaped in its surface. At the same time, this shape can be utilized to achieve the effect of ACEK. On the surface of the self-assembled electrode, there are two forms of binding antigens to antibodies.

In Figure 5, when the AC signal is not applied to the interdigital electrodes, the antigens binding to antibodies only depend on the deposition. In this case, only a small fraction of antigens have the chance to bind to the antibodies. This process takes a long time and does not guarantee the activity of antigens and antibodies. However, when an AC signal is applied to the interdigital electrodes, the ACEK effect is generated on the surface of the electrodes. Under the action of the ACEK effect, more and more antigens are rapidly enriched near the antibodies to promote the binding [38], thus improving the accuracy, rapidity, and sensitivity of detection. Hence, in this work, the ACEK effect was used on the electrodes to accelerate the binding.

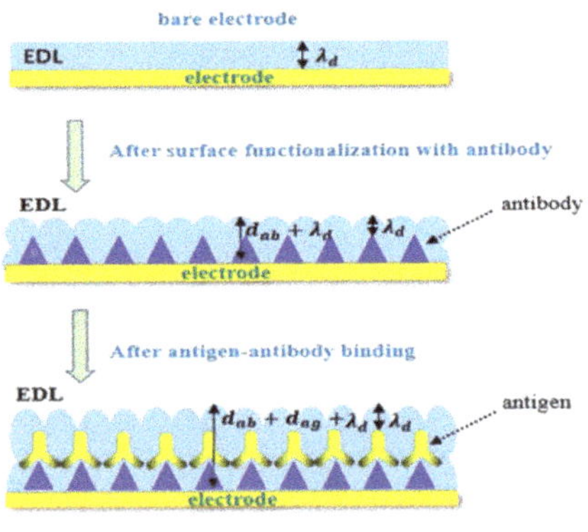

Figure 5. The performances of different detection environments on electrodes.

3. Results and Discussions

3.1. Detection of Antigen with 5.3 µg/L Antibody

In this study, different concentrations of BPA standers (1.2×10^{-7} mol/L to 1.2×10^{-4} mol/L) were tested to evaluate the performance of the method. A 10-kHz AC signal of 600 mV was applied to measure capacitance of the biosensor for 60 s.

Figure 6a shows the relationship between the normalized capacitance and different concentrations of BPA. Obviously, the change of capacitance was linear, and the rate of change of the curve increased as the concentrations of BPA increased, corresponding to the degree of binding between antibodies and antigens. The change rate of normalized capacitance curves was found by least square linear fitting, which provided a quantitative index of antibody–antigen binding. In Figure 6a, the slope of these capacitance curves was found to be −10.6‰/min, −14.6‰/min, −24.7‰/min, and −37.7‰/min for BPA levels at 1.2×10^{-7} mol/L, 1.2×10^{-6} mol/L, 1.2×10^{-5} mol/L, and 1.2×10^{-4} mol/L, respectively.

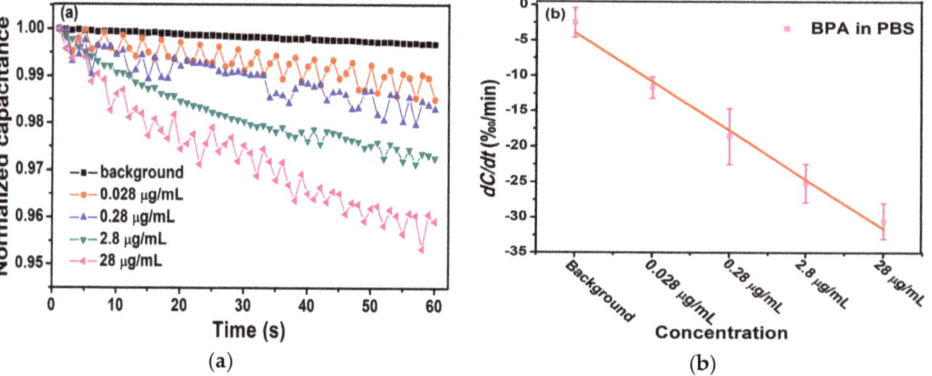

Figure 6. (**a**) Detection of different concentrations of antigen with 5.3 µg/L antibody. (**b**) The change rate of capacitance as a function of BPA concentrations in 0.01 × PBS.

In Figure 6b, we calculated the averages and standard deviations (SDs) of the biosensor response and demonstrated the correlation between the concentration of BPA and change rate of the capacitance. The range of 1.2×10^{-7} mol/L to 1.2×10^{-4} mol/L BPA samples showed change rates of −11.62‰/min ± 2.08‰/min, −18.53‰/min ± 3.93‰/min, −25.07‰/min ± 2.73‰/min, and −30.43‰/min ± 2.56‰/min, respectively. In the range of 1.2×10^{-7} mol/L to 1.2×10^{-4} mol/L BPA, *dC/dt* was logarithmically dependent on the concentration of BPA. A negative linear correlation between *dC/dt* and the concentration of BPA was observed. The dependence is expressed as $y(‰/\text{min}) = -6.912x + 3.045$ with a Pearson correlation coefficient $R^2 = 0.985$.

In the experiments, we used the impedance analyzer IM3536 to detect the change of the capacitance of the sensor when antigen–antibody binding. However, the mechanism of measurement of the impedance analyzer IM3536 is to measure the impedance and the phase angle of the device, and the capacitance is calculated from the impedance and phase angle. Considering that the developed sensor is not a pure resistor or pure capacitor, the impedance and phase angle of the sensor will change at any moment when the sensor is immersed into solution, so the capacitance has oscillation. However, the oscillation is in the range of error measurement of the impedance analyzer IM3536. In order to mitigate these oscillations in the capacitance used in the sensor transduction, it's an effective way to slow down the speed of measurement or take the average of multiple measurements.

3.2. Detection of Antigen with 5.3 μg/mL Antibody

To verify the difference in detection between electrodes modified with different antibody concentrations, 5.3 μg/mL of antibody was immobilized on the interdigital electrodes. The experimental conditions are consistent with the experiment above.

Figure 7a displays the change rate of the normalized capacitance, which was found to be −11.5‰/min, −20.0‰/min, −25.5‰/min, and −30.8‰/min for BPA concentrations at 1.2×10^{-7} mol/L, 1.2×10^{-6} mol/L, 1.2×10^{-5} mol/L, and 1.2×10^{-4} mol/L, respectively. The averages and standard deviations (SDs) of the biosensor response were exhibited in Figure 7b. From it, we can evaluate that the concentration of BPA was negatively correlated with dC/dt, and the linear correlation is expressed as $y(‰/min) = -6.93x + 1.47$ with a Pearson correlation coefficient of $R^2 = 0.989$. It follows that the electrode with a concentration of 5.3 μg/mL can also detect the BPA levels range from 1.2×10^{-7} mol/L to 1.2×10^{-4} mol/L, and the linearity and correlation are better.

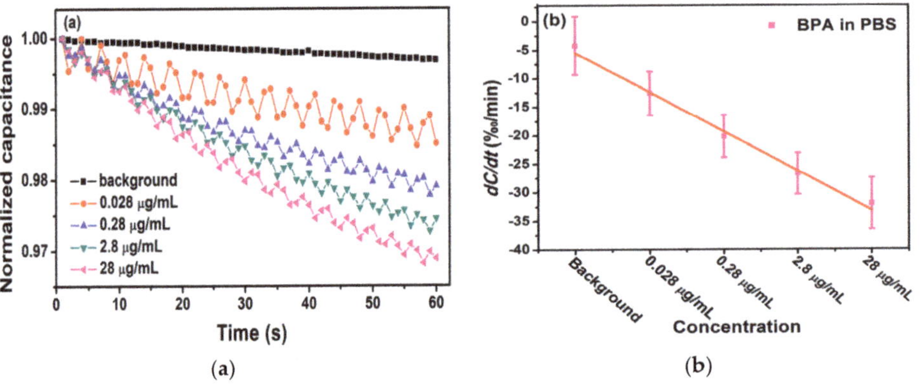

Figure 7. (a) Detection of different concentrations of antigen with 5.3 μg/mL of antibody. (b) The change rate of capacitance as a function of BPA concentrations in 0.01 × PBS.

Table 1 lists the quantitative results of the recent publications with different methods to detect BPA. Compared with most of the recent publications, which have reported the electrochemical method to achieve the detection of BPA, the results of this work may not be better than them. However, in this work, we have demonstrated a novel method to detect BPA based on self-assembly technology and the AC electrokinetics effect. There is nearly no report to detect BPA with self-assembly technology and the AC electrokinetics effect. So, it has the tremendous potential to detect BPA more sensitively with a lower limit of detection in the near future.

Table 1. Limit of detection (LOD) comparison of different methods.

Method	LOD	Reference
Carbon nanohorns/Nafion	1.8×10^{-6} mol/L	Yilin Li (2014)
SDS-(bupy)PF$_6$/CPE	3.02×10^{-7} mol/L	Zhang Yanmei (2012)
Graphene modified SPCE	4.99×10^{-7} mol/L	Ling Zhou (2014)
Strata® C18-E cartridge cleanup with detection by liquid chromatography coupled	5.21×10^{-9} mol/L	Pasquale Gallo (2019)
Nanoparticles-based fluorescence immunoassay	8.7×10^{-11} mol/L	Wei Sheng (2018)
Surface plasmon resonance (SPR) biosensor	2.28×10^{-11} mol/L	Xue C S (2019)
Gold nanoparticle-based colorimetric aptasensor	4.38×10^{-12} mol/L	Eun-Hee Lee (2019)
Electroactive label-based aptamer	3.8×10^{-10} mol/L	Maryam Nazari (2019)
A tyrosinase-based biosensor	7.7×10^{-8} mol/L	Filomeno A.D. Inroga (2019)

Table 1. *Cont.*

Method	LOD	Reference
Electrochemical enzyme biosensor	3.18×10^{-9} mol/L	Yang Liu (2019)
A signal-enhanced lateral flow strip biosensor	4.38×10^{-8} mol/L	Xiayu Peng (2017)
DNA-functionalized graphene field effect transistors integrated in microfluidic systems.	4.38×10^{-9} mol/L	Liu S (2018)
Self-assembly technology and AC electrokinetics effect	1.22×10^{-7} mol/L	This work

4. Conclusions

In this work, a rapid, highly sensitive BPA biosensor based on self-assembly technology and the AC electrokinetics (ACEK) effect has been proposed. A higher concentration of BPA solution was dropped on the self-assembly interdigital electrode sensor, a larger number of antibody–antigen binding occurred, and a larger normalized capacitance change rate was detected for the sensor. In this work, we used antigen–antibody-specific binding to detect BPA. At the same time, the limit of the biosensor is 1.2×10^{-7} mol/L, which is better than some existing detection methods. For example, Sun et al. [39] have used the oscillopolarogaphic method to detect BPA in food packaging materials with a limit of detection of 4.4×10^{-6} mol/L. Yan et al. [40] have developed a simple and renewable nanoporous gold-based electrochemical sensor for BPA detection with a limit of detection of 4.3×10^{-7} mol/L. The present work could successfully detect BPA at nanomolar (nM) levels, which was higher than the results in the previous work [32], which could successfully detect BPA at femto molar (fM) levels. Although the results of the present work were not better than the results in the previous work, we have used a novel method (self-assembly technology) to achieve the detection of BPA. At the same time, in order to acquire better results, we are improving the conditions, processes, and materials of the experiments. The novel method has the tremendous potential to detect BPA more sensitively with a lower limit of detection. Further development of the method may provide a more convenient, highly sensitive, and effective detection for BPA in complex samples.

Author Contributions: Methodology, formal analysis, investigation, data curation, writing—original draft, H.L., Z.P.; conceptualization, validation, writing—review and editing, supervision, project administration, funding acquisition, X.L.; resources, writing—original draft, M.S., L.J. All authors have read and agreed to the published version of the manuscript.

Funding: This work was supported by the Fundamental Research Funds for the Central Universities (Project No.2019CDYGZD006), National Natural Science Foundation of China (Project no. 6137 7001), the Postgraduate education and teaching reform research project of Chongqing University (Project No. cquyjg18323), Venture & Innovation Support Program for Chongqing Overseas Returnees and Fundamental Research Funds for the Central Universities (No. 10611CDJXZ238826).

Acknowledgments: The authors thank Xiaodong Zheng for his support in validation and writing—review and editing.

Conflicts of Interest: The authors declare no conflict of interest. The funders had no role in the design of the study; in the collection, analyses, or interpretation of data; in the writing of the manuscript, or in the decision to publish the results.

References

1. Chevalier, N.; Brucker-Davis, F.; Lahlou, N.; Coquillard, P.; Pugeat, M.; Pacini, P.; Panaïa-Ferrari, P.; Wagner-Mahler, K.; Fénichel, P. A negative correlation between insulin-like peptide 3 and bisphenol A in human cord blood suggests an effect of endocrine disruptors on testicular descent during fetal development. *Hum. Reprod.* **2014**, *30*, 447–453. [CrossRef] [PubMed]
2. Lang, I.A.; Galloway, T.S.; Scarlett, A.; Henley, W.E.; Depledge, M.; Wallace, R.B.; Melzer, D. Association of Urinary Bisphenol A Concentration With Medical Disorders and Laboratory Abnormalities in Adults. *JAMA* **2008**, *300*, 1303–1310. [CrossRef] [PubMed]

3.	Mei, Z.; Qu, W.; Deng, Y.; Chu, H.; Cao, J.; Xue, F.; Zheng, L.; El-Nezamic, H.S.; Wu, Y.; Chen, W. One-step signal amplified lateral flow strip biosensor for ultrasensitive and on-site detection of bisphenol A (BPA) in aqueous samples. *Biosens. Bioelectron.* **2013**, *49*, 457–461. [CrossRef] [PubMed]

4.	Apodaca, D.C.; Pernites, R.B.; Ponnapati, R.; Del Mundo, F.R.; Advincula, R.C. Electropolymerized Molecularly Imprinted Polymer Film: EIS Sensing of Bisphenol A. *Macromol.* **2011**, *44*, 6669–6682. [CrossRef]

5.	Soto, A.M.; Sonnenschein, C. Environmental causes of cancer: Endocrine disruptors as carcinogens. *Nat. Rev. Endocrinol.* **2010**, *6*, 363–370. [CrossRef]

6.	D'Antuono, A.; Dall'Orto, V.C.; Balbo, A.L.; Sobral, S.; Rezzano, I. Determination of bisphenol A in food-simulating liquids using LCED with a chemically modified electrode. *J. Agric. Food Chem.* **2001**, *49*, 1098–1101. [CrossRef]

7.	Nakazawa, H.; Yamaguchi, A.; Inoue, K.; Yamazaki, T.; Kato, K.; Yoshimura, Y.; Makino, T. In vitro assay of hydrolysis and chlorohydroxy derivatives of bisphenol A diglycidyl ether for estrogenic activity. *Food Chem. Toxicol.* **2002**, *40*, 1827–1832. [CrossRef]

8.	Fontana, A.R.; De Toro, M.M.; Altamirano, J.C. One-Step Derivatization and Preconcentration Microextraction Technique for Determination of Bisphenol A in Beverage Samples by Gas Chromatography–Mass Spectrometry. *J. Agric. Food Chem.* **2011**, *59*, 3559–3565. [CrossRef]

9.	Yin, W.; Wu, L.; Ding, F.; Li, Q.; Wang, P.; Li, J.; Lu, Z.; Han, H. Surface-imprinted SiO2@Ag nanoparticles for the selective detection of BPA using surface enhanced Raman scattering. *Sensors Actuators B Chem.* **2018**, *258*, 566–573. [CrossRef]

10.	Kubo, I.; Kanamatsu, T.; Furutani, S. Microfluidic device for enzyme-linked immunosorbent assay (ELISA) and its application to bisphenol a sensing Sens. *Mater* **2014**, *26*, 615–621.

11.	Gallo, P.; Pisciottano, I.D.M.; Fattore, M.; Rimoli, M.G.; Seccia, S.; Albrizio, S. A method to determine BPA, BPB, and BPF levels in fruit juices by liquid chromatography coupled to tandem mass spectrometry. *Food Addit. Contam. Part A* **2019**, *36*, 1871–1881. [CrossRef] [PubMed]

12.	Sheng, W.; Duan, W.; Shi, Y.; Chang, Q.; Zhang, Y.; Lu, Y.; Wang, S. Sensitive detection of bisphenol A in drinking water and river water using an upconversion nanoparticles-based fluorescence immunoassay in combination with magnetic separation. *Anal. Methods* **2018**, *10*, 5313–5320. [CrossRef]

13.	Xue, C.S.; Erika, G.; Jiří, H. Surface plasmon resonance biosensor for the ultrasensitive detection of bisphenol A. *Anal. Bioanal. Chem.* **2019**, *411*, 5655–5658. [CrossRef] [PubMed]

14.	Avila, M.; Zougagh, M.; Ríos, Á.; Escarpa, A. Molecularly imprinted polymers for selective piezoelectric sensing of small molecules. *TrAC Trends Anal. Chem.* **2008**, *27*, 54–65. [CrossRef]

15.	Brusnitsyn, D.V.; Medyantseva, E.P.; Varlamova, R.M.; Sitdikova, R.R.; Fattakhova, A.N.; Konovalova, O.A.; Budnikov, G.K. Carbon nanomaterials as electrode surface modifiers in development of amperometric monoamino oxidase biosensors. *Inorg. Mater.* **2016**, *52*, 1413–1419. [CrossRef]

16.	Vashist, S.K.; Zheng, D.; Al-Rubeaan, K.; Luong, J.H.; Sheu, F.-S. Advances in carbon nanotube based electrochemical sensors for bioanalytical applications. *Biotechnol. Adv.* **2011**, *29*, 169–188. [CrossRef]

17.	Ntsendwana, B.; Sampath, S.; Mamba, B.; Arotiba, O. Photoelectrochemical oxidation of p-nitrophenol on an expanded graphite–TiO2 electrode. *Photochem. Photobiol. Sci.* **2013**, *12*, 1091. [CrossRef]

18.	Ndlovu, T.; Arotiba, O.; Sampath, S.; Krause, R.; Mamba, B. Electroanalysis of copper as a heavy metal pollutant in water using cobalt oxide modified exfoliated graphite electrode. *Phys. Chem. Earth Parts A/B/C* **2012**, *50*, 127–131. [CrossRef]

19.	Liu, X.; Peng, Y.; Qu, X.; Ai, S.; Han, R.; Zhu, X. Multi-walled carbon nanotube-chitosan/poly (amidoamine)/DNA nanocomposite modified gold electrode for determination of dopamine and uric acid under coexistence of ascorbic acid. *J. Electroanal. Chem.* **2011**, *654*, 72–78. [CrossRef]

20.	Manivel, P.; Dhakshnamoorthy, M.; Balamurugan, A.; Ponpandian, N.; Mangalaraj, D.; Viswanathan, C.; N, P. Conducting polyaniline-graphene oxide fibrous nanocomposites: Preparation, characterization and simultaneous electrochemical detection of ascorbic acid, dopamine and uric acid. *RSC Adv.* **2013**, *3*, 14428. [CrossRef]

21.	Li, Z.; Nambiar, S.; Zheng, W.; Yeow, J. PDMS/single-walled carbon nanotube composite for proton radiation shielding in space applications. *Mater. Lett.* **2013**, *108*, 79–83. [CrossRef]

22.	Nazari, M.; Kashanian, S.; Rafipour, R.; Omidfar, K. Biosensor design using an electroactive label-based aptamer to detect bisphenol A in serum samples. *J. Biosci.* **2019**, *44*, 105. [CrossRef] [PubMed]

23. Lee, E.-H.; Lee, S.K.; Kim, M.J.; Lee, S.-W. Simple and rapid detection of bisphenol A using a gold nanoparticle-based colorimetric aptasensor. *Food Chem.* **2019**, *287*, 205–213. [CrossRef]
24. Inroga, F.A.D.; Rocha, M.O.; Lavayen, V.; Arguello, J. Development of a tyrosinase-based biosensor for bisphenol A detection using gold leaf–like microstructures. *J. Solid State Electrochem.* **2019**, *23*, 1659–1666. [CrossRef]
25. Liu, Y.; Yao, L.; He, L.; Liu, N.; Piao, Y. Electrochemical Enzyme Biosensor Bearing Biochar Nanoparticle as Signal Enhancer for Bisphenol A Detection in Water. *Sensors* **2019**, *19*, 1619. [CrossRef] [PubMed]
26. Peng, X.; Kang, L.; Pang, F.; Li, H.; Luo, R.; Luo, X.; Sun, F. A signal-enhanced lateral flow strip biosensor for ultrasensitive and on-site detection of bisphenol A. *Food Agric. Immunol.* **2018**, *29*, 216–227. [CrossRef]
27. Liu, S.; Fu, Y.; Xiong, C.; Liu, Z.; Zheng, L.; Yan, F. Detection of Bisphenol A Using DNA-Functionalized Graphene Field Effect Transistors Integrated in Microfluidic Systems. *ACS Appl. Mater. Interfaces* **2018**, *10*, 23522–23528. [CrossRef]
28. Li, S.; Cui, H.; Yuan, Q.; Wu, J.; Wadhwa, A.; Eda, S.; Jiang, H. AC electrokinetics-enhanced capacitive immunosensor for point-of-care serodiagnosis of infectious diseases. *Biosens. Bioelectron.* **2014**, *51*, 437–443. [CrossRef]
29. Liu, X.; Yang, K.; Wadhwa, A.; Eda, S.; Li, S.; Wu, J. Development of an AC electrokinetics-based immunoassay system for on-site serodiagnosis of infectious diseases. *Sens. Actuators A Phys.* **2011**, *171*, 406–413. [CrossRef]
30. Lian, M.; Islam, N.; Wu, J. AC electrothermal manipulation of conductive fluids and particles for lab-chip applications. *IET Nanobiotechnology* **2007**, *1*, 36–43. [CrossRef]
31. Cheng, C.; Cui, H.; Wu, J.; Eda, S. A PCR-free point-of-care capacitive immunoassay for influenza A virus. *Microchim. Acta* **2017**, *55*, 53–1657. [CrossRef]
32. Lin, X.; Cheng, C.; Terry, P.; Chen, J.; Cui, H.; Wu, J. Rapid and sensitive detection of bisphenol a from serum matrix. *Biosens. Bioelectron.* **2017**, *91*, 104–109. [CrossRef] [PubMed]
33. Bertilsson, L.; Potje-Kamloth, K. Molecular interaction of DMMP and water vapor with mixed self-assembled mon0layers studied by IR spectroscopy and SAW devices. *7rhin Solid Film.* **1996**, *284*, 882–887. [CrossRef]
34. Yang, X.; Shi, J.; Johnson, S.; Swanson, B. Molecular host siloxane thin films for surface acoustic wave chemical sensors. *Sens. Actuators B Chem.* **1997**, *45*, 79–84. [CrossRef]
35. Helmholtz. On the modern development of Faraday's conception of electricity. *J. Franklin Inst.* **1881**, *44*, 182–185.
36. Grahame, D.C. The electrical double layer and the theory of electrocapillarity. *Chem. Rev.* **1947**, *41*, 441–501. [CrossRef]
37. Berggren, C.; Bjarnason, B.; Johansson, G. Capacitive biosensors. *Electroanalysis* **2001**, *13*, 173–180. [CrossRef]
38. Wu, J. Interactions of electrical fields with fluids: Laboratory-on-a-chip applications. *IET Nanobiotechnology* **2008**, *2*, 14. [CrossRef]
39. Shi-Ping, S.; Zhi-Dong, M.A.; Wen-De, Z. Determination of Bisphenol A in Food Packaging Materials by Oscillopolarogaphic Method. *J. Anal. Sci.* **2002**, *18*, 490–492.
40. Yan, X.; Zhou, C.; Yan, Y.; Zhu, Y. A Simple and Renewable Nanoporous Gold-based Electrochemical Sensor for Bisphenol A Detection. *Electroanal.* **2015**, *27*, 2718–2724. [CrossRef]

Review

Advances in Liquid Metal-Enabled Flexible and Wearable Sensors

Yi Ren [1], Xuyang Sun [2] and Jing Liu [1,2,*]

[1] Department of Biomedical Engineering, School of Medicine, Tsinghua University, Beijing 100084, China; ren-y18@mails.tsinghua.edu.cn

[2] Beijing Key Lab of CryoBiomedical Engineering and Key Lab of Cryogenics, Technical Institute of Physics and Chemistry, Chinese Academy of Sciences, Beijing 100190, China; sunxuy@mail.ipc.ac.cn

* Correspondence: jliubme@tsinghua.edu.cn; Tel.: 86-10-62794896

Received: 8 January 2020; Accepted: 13 February 2020; Published: 15 February 2020

Abstract: Sensors are core elements to directly obtain information from surrounding objects for further detecting, judging and controlling purposes. With the rapid development of soft electronics, flexible sensors have made considerable progress, and can better fit the objects to detect and, thus respond to changes more sensitively. Recently, as a newly emerging electronic ink, liquid metal is being increasingly investigated to realize various electronic elements, especially soft ones. Compared to conventional soft sensors, the introduction of liquid metal shows rather unique advantages. Due to excellent flexibility and conductivity, liquid-metal soft sensors present high enhancement in sensitivity and precision, thus producing many profound applications. So far, a series of flexible and wearable sensors based on liquid metal have been designed and tested. Their applications have also witnessed a growing exploration in biomedical areas, including health-monitoring, electronic skin, wearable devices and intelligent robots etc. This article presents a systematic review of the typical progress of liquid metal-enabled soft sensors, including material innovations, fabrication strategies, fundamental principles, representative application examples, and so on. The perspectives of liquid-metal soft sensors is finally interpreted to conclude the future challenges and opportunities.

Keywords: soft sensors; liquid metal; fabrication; principle; arrays; application

1. Introduction

Sensors are common and core devices for information acquisition, which can detect external stimuli and transform them into measurable signals according to specific rules. In particular, biomedical sensors are targeted at human beings and other biological organisms to monitor physiological signals of life activity for health monitoring and disease diagnosis [1].

A typical sensor mainly consists of three components, including sensing element, conversion element and electrical circuits. The sensing element directly perceives the measured quantity and converts the signal into other values following a definite relationship, while the conversion element converts the non-electric quantity sensed by the sensitive element into electrical quantity [2]. Generally, the sensitive element and conversion unit are integrated as a whole. Diverse sensing materials are utilized to sense environmental change and produce a useful signal, such as piezoelectric effect, thermoelectric effect, Hall Effect etc. [2]. According to various detecting principles, existing sensors mainly include pressure sensors, force sensors, liquid sensors, speed sensors, acceleration sensors, radiation sensors, thermal sensors and so on [3–9].

With the rapid development of soft electronics [10–20], flexible and wearable sensors have shown evident advantages in biomedical applications, such as health-monitoring, wearable devices, artificial skin and intelligent robotics [10,21–31]. For instance, strain sensors can be attached to human body to monitor skin strain and muscle movement [32]. Introduction of soft sensors has greatly propelled the development of intelligent artificial skin, which provides not only aesthetic functions, but also perception of tactile sensation and temperature measurement [33,34]. They can also be applied to realize intelligent perception and control for robots [35–37]. Moreover, flexible sensors can tightly fit the object and maintain performance even while being stretched. In this way, they can respond to changes more sensitively and detect position and posture alteration in real time.

Up to now, materials and techniques to achieve flexible sensors have been widely explored. To obtain better application output for the human body, both flexibility and biocompatibility of the substrates are required. Soft thermoplastic polymers and silicone elastomers are the most common substrates used to fabricate flexible sensors, such as polyethylene terephthalate (PET), poly urethane (PU), polydimethylsiloxane (PDMS) and Eco Flex [38]. Moreover, the sensing elements are other important parts, which are mainly made up of conductive materials, including organic and inorganic nanomaterials [39,40]. Carbon-based materials, such as carbon nanotube and graphene, have been investigated in various soft circuits [39,41–46]. Although circuits fabricated by carbon-based materials are highly stretchable, they are not as conductive as those fabricated by metals. Nanoparticles of rigid metals are applied to design soft circuits with high conductivity, among which silver nanoparticles have been extensively studied [47–53]. Moreover, liquid metal, mainly refers to metal whose melting point is around the room temperature, is increasingly popular in the area of soft circuits because it possesses both good deformability and conductivity [54]. The room-temperature melting alloys can be fabricated by stirring in a liquid state, and sometimes heating is needed. For instance, the conductivity of gallium is $2.2 \times 10^6 \ S/m$, which is a little lower than that of silver but is much higher than that of organic materials [54]. Furthermore, liquid metal is more flexible and cheaper than silver. Thus, progress in liquid metal has been expected to propel the development of soft circuits. Typical liquid metal includes gallium, bismuth, lead, tin, cadmium, indium and their alloys. These low-melting alloys can be produced by stirring in a liquid state, and some raw materials with high melting points need to be heated. When exposed to the air, liquid metal will be oxidized immediately and form a layer of protective film. Unlike conventional liquid metal like mercury, gallium and its alloys is proven to be low-toxic and biocompatible, and has been introduced to biomedical area [55].

Due to the unique properties of liquid metal, it has been vigorously studied in soft electronics, such as flexible conductors, antennas, transistors, electrodes, energy-harvesting devices, sensors, liquid robots and liquid computational systems [56–68]. In particular, liquid-metal sensors are designed on the basis of variable resistance and capacitance during stretching and bending. Compared to nanoparticles of rigid metals, liquid metal appears more fluidic and deformable. Therefore, liquid-metal based sensors would fit well to the targeted area, further reducing discomfort and displaying better sensitivity and flexibility. The main problem that must be paid attention to liquid metal during application is the risk of leaking from soft encapsulation.

In this review, the progress of soft sensors based on liquid metal is reviewed. First, we introduce the composition of liquid metal, as well as its distinct properties thus enabled. Next, fabrication methods of liquid-metal sensors and general comparisons among them are discussed. Subsequently, we move to interpret the basic principles of how liquid-metal sensors work. Following that, several typical sensors or sensor arrays based on liquid metal and their representative applications are illustrated. Finally, we point out the existing issues that would hinder further development in this area, and propose possible strategies for future research.

2. Materials

2.1. Composition

At this stage, liquid metal applied to soft electronics mainly refers to gallium (Figure 1A) and its alloys, among which EGaIn (gallium and indium) and GaInstan (gallium, indium and tin) are most commonly available [55]. EGaIn contains 75.5% gallium and 24.5 % indium by weight and its melting point is 15.5 °C, while Galinstan consists of 68.5% gallium, 21.5% indium and 10.0% tin by weight and its melting point is 10.5 °C. Those alloys can be prepared by stirring in liquid state and sometimes heating is required. Similar to gallium, bismuth can be mixed up with other metals to form low-melting alloys as well. Their physical properties like melting point and electric conductivity can be regulated according to the addition of different metals [69].

In particular, the properties of liquid metal can be optimized by mixing liquid metal with other metal nanoparticles. The doping ratio decides the fluidity of the mixture. When the ratio of metal nanoparticles is high enough, the mixture will be in the form of a paste. For instance, magnetism, ferromagnetic nano powders could be dispersed in liquid metal [55]. Moreover, nickel particles could be mixed with EGaIn for fast fabrication of soft electronics by direct writing or printing [70–72]. As shown in Figure 1B, Ni particles were wrapped by a liquid metal oxide layer after stirring for a moment. The introduction of Ni particles could help enhance the adhesion force of liquid metal to substrate. Furthermore, to improve its conductivity, one could load copper particles into liquid metal to make ever superior composite [73]. Figure 1C shows the mixture of Cu-GaIn. The doping ratio and particle diameter could be adjusted according to different application requirements. Thermal and mechanical properties were tunable as well.

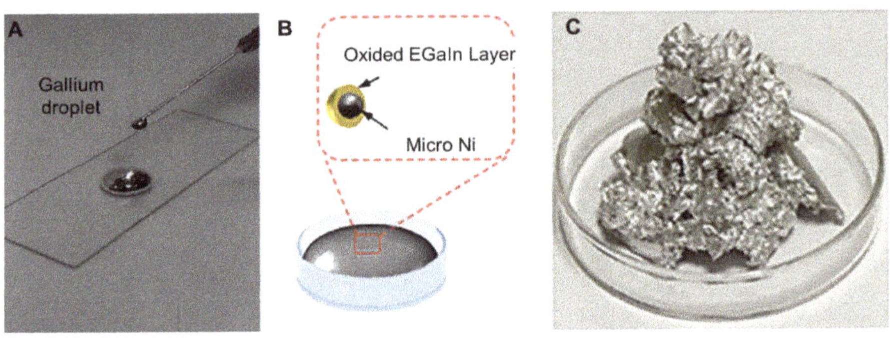

Figure 1. Liquid metal and its mixture. (**A**) Photo of gallium droplet, reproduced with permission from [69]. (**B**) Schematic diagram of Ni-GaIn particle, reproduced with permission from [70]. (**C**) Photo of Cu-GaIn mixure, reproduced with permission from [73].

2.2. Property

It is well known that liquid metal exhibits unique physical and chemical properties, which have been extensively studied. The most outstanding feature of liquid metal is that it has both fluidity and metallicity, making it advantageous to soft electronics. Previous experiments have testified that although liquid metal is fluidic, its dynamic performance is different from that of water, such as droplet shape and splashing morphology [74]. Figure 2A shows the splashing process of water and liquid metal under 25 °C. Clearly, it is hard for liquid metal to form secondary droplets due to high dynamic viscosity, which is really easy for splashing water. Moreover, gallium and its alloys could be oxidized quickly when exposed to air, which would lead to an even higher viscosity, increasing the difficulty to form secondary droplets. But on the other hand, formation of the oxide skin can increase the adhesive force, thus providing convenience for fabrication of liquid metal circuits. Research has found that liquid metal possesses a high surface tension and is hardly permeable on various substrate materials [75]. Metal particles can be mixed to regulate the adhesion and wettability of liquid metal, as mentioned above.

Liquid metal shows good electrical and thermal conductivity. The electrical conductivity of EGaIn is around 10^6 S/m and the thermal conductivity is much higher than that of water. Moreover, when applying an electric or magnetic field, liquid metal would respond [76]. Coated with ferromagnetic materials, liquid-metal manipulation could be induced by an external magnetic field. As shown in

Figure 2B, liquid-metal droplets could move following the magnet. An electrical field could induce the movement of liquid metal as well [77]. Figure 2C presents the movement of an EGaIn droplet under the control of an electrical filed. Furthermore, Tan et al. studied the movement of liquid-metal droplets with and without aluminum, respectively, and found that when there existed aluminum, liquid-metal droplets could run faster under the control of an electric field [78]. This is because that EGaIn and aluminum would constitute galvanic battery in electrolyte solution and the gas generated could propel the movement of the liquid metal. This redox can be represented by the chemical formula below:

$$2Al + 6H^+ \rightarrow 2Al^{3+} + 3H_2. \tag{1}$$

Since liquid metal shows these distinctive properties, the application of liquid metal in soft circuits is attracting increasing attention.

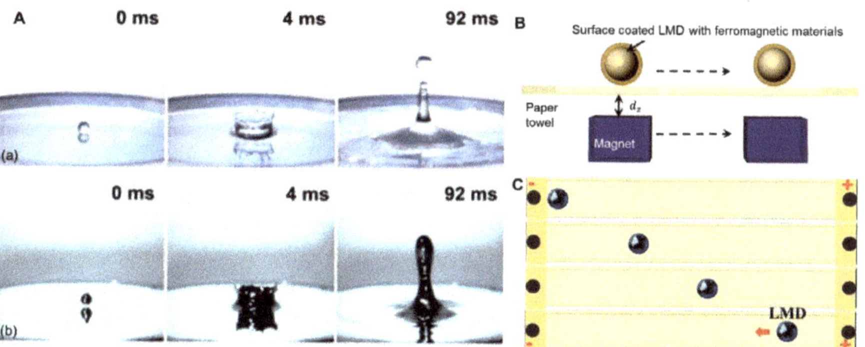

Figure 2. Typical properties of liquid metal. (**A**) Comparison of splashing process of water (a) and liquid metal (b), reproduced with permission from [74]; (**B**) Schematic of the magnet-controlled moving of an EGaIn droplet coated by ferromagnetic materials, reproduced with permission from [76]; (**C**) Sequential snapshots of a EGaIn droplet's motion, reproduced with permission from [76];.

3. Fabrication

Owing to the unique physical properties of liquid metal, different methods to fabricate liquid-metal soft circuits have been designed and explored. Up to now, printing and microfluidic technologies are most commonly used [79]. Other methods including selective wetting, laser engraving, wiping etc., are investigated as well.

3.1. Printing Technology

Based on the fluidity of liquid metal, the concept of liquid-metal printed electronics was proposed [79,80]. Researches have proven that liquid metal can be used as ink and printed directly by a brush or a roller ball pen [81–83]. Compared to a traditional printed circuit board, this made electronics fabrication more convenient and rapid. Figure 3A shows how to paint a circuit with a roller ball pen and the light-emitting diodes (LEDs) are successfully lighted. To standardize the fabrication process of soft circuits by liquid metal ink and avoid artificial errors, automatic printing of liquid metal by a versatile desktop printer was further developed [84,85]. The concrete structure of a desktop printer with liquid metal ink is presented in Figure 3B. Circuits could be precisely designed by the desktop printer and then transferred to soft substrate, such as PDMS. Those circuits would be further encapsulated for practical application [86]. Votzke et al. designed a highly stretchable strain sensor using printed liquid metal, which has been used to measure elbow flexion angle [87]. Furthermore, Gannarapu et al. proposed the method of freeze-printing to fabricate 3D curvilinear paths. Liquid metal was continuously dispensed and frozen. The printed solid network could be encapsulated in elastomers [88]. A shortage of direct printing lies in that the resolution of circuits is sometimes limited by the resolution of the printer and can hardly be improved. To design liquid metal with smaller size, Liang et al. wrote liquid metal on a prestretched elastomeric substrate surface [89]. Then the prestretched substrate was released to recover the original size with electronic patterns shrinking on it.

Recently, mask-based printing has been designed as an alternative to promote resolution. For instance, a rigid mask could be fabricated by micromachining technology in advance. Areas covered by the mask would not be available to liquid metal and specific patterns could be deposited through the mask with an airbrush (Figure 3C) [90]. Moreover, a photoresist mask was commonly used as well, which can be removed finally, as well as the liquid metal falling on it [91,92]. However, unlike a rigid mask, a photoresist mask cannot be recycled and the process of photoetching is complex and time-consuming. Guo et al. introduced toner as the mask and liquid metal could selectively adhere to area without toner [72]. A roller was applied to deposit liquid metal and the circuit could further be transferred to elastic substrate. Figure 3D shows the whole process of Guo et al.'s method.

In general, compared to other methods, printing technology may be more convenient, but the fabrication precision is relatively limited.

3.2. Microfluidic Technology

Meanwhile, microfluidic technology is advantageous for the fabrication of stable, uniform and sealed liquid-metal circuits [93–99]. Figure 3E is the schematic of the microfluidic technique. Firstly, photoetching was applied to produce a raised pattern model. Then, the mixture of PDMS prepolymer and curing agent with the quantity ratio of 10:1 was poured onto the model to form a PDMS layer with specific grooves. Plasma bonding was utilized to form sealed micro-channels. To fill the channels with liquid metal and exhaust air, a pair of inlet and outlet holes must be produced before plasma bonding. After that, liquid metal could be injected into the channels. Circuits fabricated by microfluidic technology showed excellent stability and uniformity. Microfluidic technology has already been applied widely and commercially.

Several types of soft force sensors have, thus, been designed by this method. As shown in Figure 3F, the soft pressure sensor was fabricated by microfluidic technology. Research has proven that the soft force sensors could generate a response to external pressure through the change of capacitance and exhibit good flexibility.

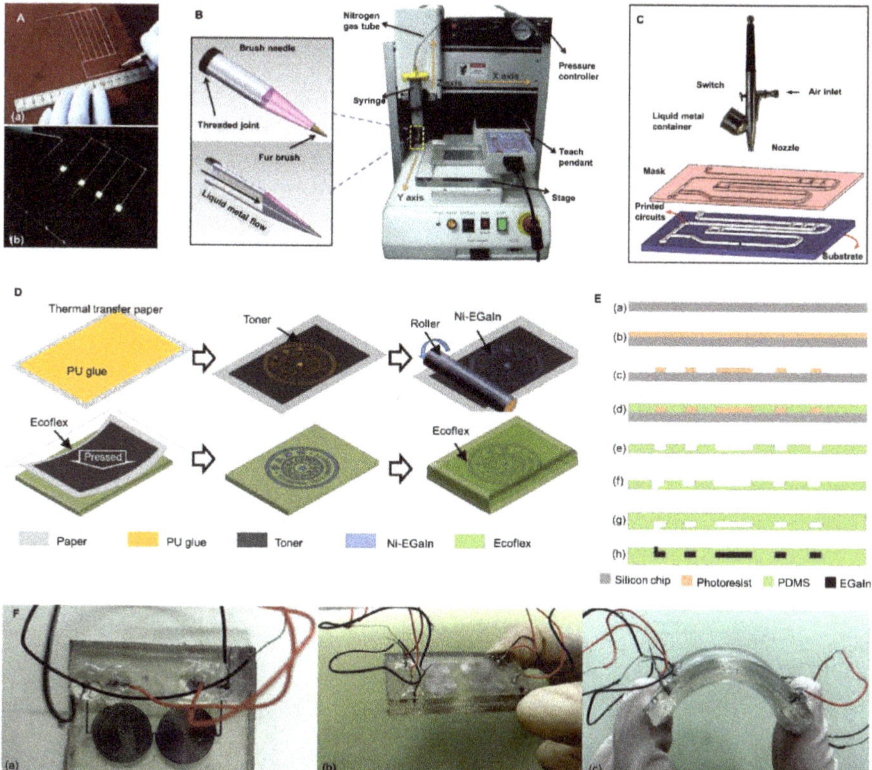

Figure 3. Printing and microfluidic technology. (**A**) Printing circuits with liquid metal ink by a roller ball pen, reproduced with permission from [82]. (**B**) Mechanical printing of liquid metal circuits with desktop printer, reproduced with permission from [84]. (**C**) Schematic of patterning liquid metal with a rigid mask, reproduced with permission from [90]. (**D**) Schematic of patterning liquid metal with toner mask, reproduced with permission from [72]. (**E**) Schematic of microfluidic technology: (**a**) The silicon chip; (**b**) Silicon chip coated by photoresist; (**c**) Lithography model; (**d**) Pouring PDMS; (**e**) PDMS with micro-grooves; (**f**) Punching; (**g**) Plasma bonding; (**h**) Injecting liquid metal. (**F**) A stretchable pressure sensor fabricated by the microfluidic technology: (**a**) Vertical view of the pressure sensor; (**b**) Original state of the pressure sensor; (**c**) A pressure sensor under bending; Reproduced with permission from [96].

3.3. Selective Wetting

Not limited to printing and microfluidic technology, fabrication methods of flexible sensors based on liquid metal are diverse due to the unique properties of liquid metal [79]. For instance, considering that the wetting behavior of liquid metal varies with the surface properties of the substrate, selective wetting of liquid metal is studied.

Kramer et al. had fabricated elastomer conductors by deposition, which was based on the different wetting behavior between liquid metal and thin metal films [100]. Figure 4A shows the whole process of fabricating circuits by this method. The surface was sputtered with indium, which prevented liquid metal from adhering to this area. As shown in Figure 4B, the circuits thus fabricated on a soft substrate were proven to be highly flexible. Similarly, Li et al. presented a stretchable pulse sensor fabricated by selective plating as well, in which a thin film of copper would be electroplated in advance to enhance the wetting of liquid metal [101]. Clearly, surface modification could be utilized by either enhancing or weakening the wettability of liquid metal so as to form designed patterns. The limitation of this method may be that the process of surface modification is complex and the edge of the pattern produced by selective wetting is rough.

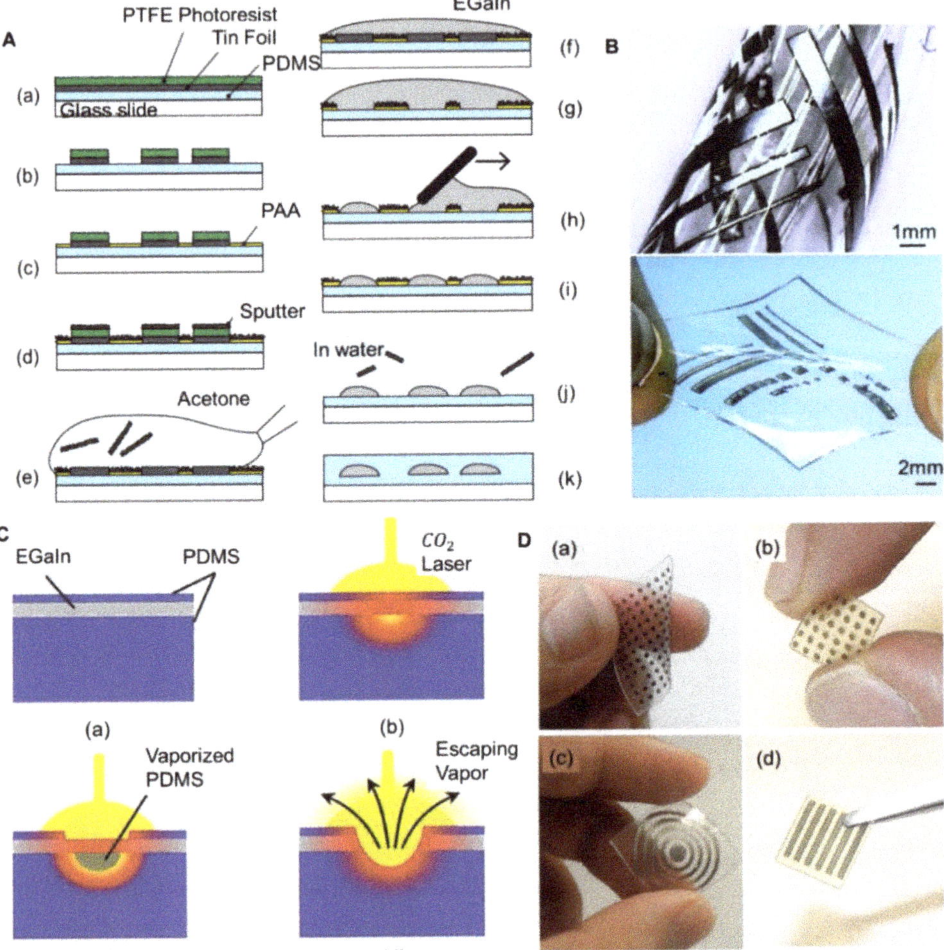

Figure 4. Diverse techniques to fabricate liquid-metal soft electronics. (**A**) Fabrication process of selective liquid metal deposition: (**a**) Photoetching; (**b**) Photoresist developing; (**c**) Spinning poly acrylic acid (PAA); (**d**) Sputtering the surface with indium; (**e**) Removing photoresist; (**f,g**) Flooding the surface with liquid metal; (**h,i**) Removing excess liquid metal; (**j**) Removing indium mask; (**k**) Encapsulating the circuit. Reproduced with permission from [100]. (**B**) A circuit embedded in elastomer patterned by selective deposition of liquid metal, reproduced with permission from [100]. (**C**) Illustration of CO_2 laser engraving technology, reproduced with permission from [102]. (**D**) Flexible circuits fabricated by CO_2 laser, reproduced with permission from [102].

3.4. Laser Engraving

The laser engraving technique is commonly applied in the field of micromachining. CO_2 laser engraving technology was introduced for rapid prototyping of soft electronics as well [102]. Figure 4C shows the direct patterning of liquid metal by a CO_2 laser cutter. Firstly, a thin layer of liquid metal was sealed by two layers of PDMS and the top layer mainly functioned as protection during the laser-cutting process. PDMS could absorb the energy from the CO_2 laser and be vaporized. Then liquid metal would escape together with PDMS vapor, thus forming desired patterns. After that, a new layer of PDMS would be cast to encase the circuit. Lu et al. have designed several systems through this technology, including sensors and resistive circuits (Figure 4D). CO_2 laser-engraving technology is advantageous for the fabrication of any 2D structure. However, it causes a higher wastage of liquid metal since a large proportion of liquid metal escaped with vaporized PDMS, which could be avoided by collecting the liquid metal that escaped. Recently, NdYAG lasers were applied for simpler patterning of liquid metal [103]. Pan et al. produced a film composed of a liquid metal grid through direct laser writing, and proved that the circuit showed high optical transmittance, high stretchability, low resistivity and low trace width.

3.5. Dewetting and Wiping

Meanwhile, methods of dewetting [104] and wiping [105] utilized the property that liquid metal would be oxidized immediately and form a thin oxide layer when exposed to the air or oxygen. The oxide layer could increase the adhesion between liquid metal and micro-channels. Therefore, while liquid metal outside the channels was removed, liquid metal in the channels would be left owing to the higher adhesion force. The difference is that dewetting uses NaOH solution to remove excess liquid metal and wiping uses absolute ethyl alcohol (Figure 5A). Compared to dewetting by NaOH solution, wiping by absolute ethyl alcohol is more controllable since the reaction between NaOH solution and liquid metal is more intense.

As shown in Figure 5A, the first few steps of wiping were similar to microfluidic technology: photoetching was used to form the desired pattern model and micro-grooves. But inlet and outlet holes were unnecessary, which significantly simplified the fabrication process and improved the resolution. After spraying liquid metal onto PDMS by a nozzle, paper soaked with absolute alcohol was applied to remove excess liquid metal. The circuits could be encapsulated by casting another layer of PDMS. Figure 5B (a) is an electrode array with high resolution fabricated by spraying and wiping. Figure 5B (b),(c) show the detailed micrograph of the electrode and lines. It could be concluded that the channels were fully filled by liquid metal. A higher failure rate may be the main limitation of the wiping technique, but this can be made up by increasing the depth of micro-channels and repeating the operations of spraying and wiping.

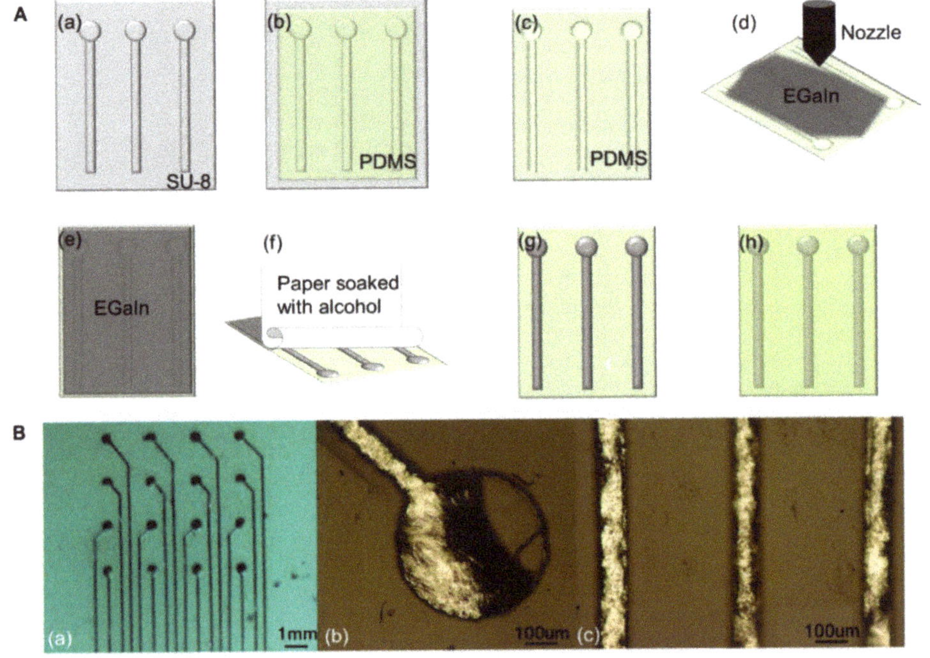

Figure 5. Technologies of spraying and wiping. (**A**) Fabrication process: (**a**) Photolithography model; (**b**) Pouring PDMS onto the photolithography model; (**c**) PDMS substrate with micro-channels; (**d,e**) Spraying liquid metal with an airbrush; (**f**) Removing needless liquid metal with the help of paper soaked with absolute ethyl alcohol; (**g**) The final circuit; (**h**) Encapsulating the circuit. (**B**) An electrode array fabricated by spraying and wiping: (**a**) The electrode array; (**b**) The micrograph of the electrode plate; (**c**) The micrograph of lines. Reproduced with permission from [105].

4. Basic Principle of Liquid-Metal Sensors

4.1. Liquid Metal as Soft Connection

The simplest liquid-metal sensors are perhaps those that use liquid metal as soft connections (Figure 6). The detecting mechanism is the same as rigid ones, but the stretchability is greatly improved. Liquid metal possesses both good conductivity and fluidity, ensuring the reliability of the connections during stretching and bending. For instance, based on photoplethysmography (PPG) technology, Li et al. designed a stretchable pulse biosensor with liquid-metal interconnections and soft substrate [101]. This system was proven to be convenient and comfortable for heart-rate monitoring. In addition, Hong et al. designed a soft polyaniline nanofiber temperature sensor array, whose interconnections were made up of Galinstan [106]. It could be used as a portable and wearable device to measure body temperature. Combining liquid-metal wiring with graphene sensors, Jiao et al. developed several strain sensors, which could be used for health-monitoring and motion-sensing [107]. Meanwhile, Hu et al. put forward the idea of using liquid metal to connect the sensing element and successfully produced a sensor array, which had the ability to sense temperature and force [108]. All these research works prove that a liquid-metal connection provides advantages in fabricating flexible sensors.

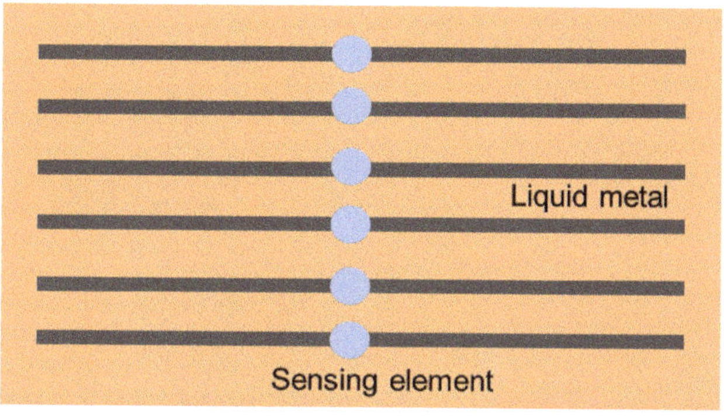

Figure 6. Illustration of circuits using liquid metal as soft connection.

4.2. Resistive Sensors

Resistive sensors based on liquid metal have been widely researched. External change can be reflected through the fluctuations in resistance. The resistance R follows the formula:

$$R = \rho L/(S),\tag{2}$$

where ρ is the resistivity of liquid metal; L is the length and S is the cross-sectional area. External heat and force can lead to the deformation of liquid metal, including thermal expansion, stretching and pressing, which further causes a change in resistance. As shown in Figure 7A, stretching can influence the length and cross-sectional area of liquid-metal lines, while pressing can change the cross-sectional area. When being bent, the shape of liquid metal does not change significantly, as well as the resistance.

Generally, liquid-metal resistive sensors include temperature sensors and force sensors. For instance, Li et al. has found the matching metal of liquid metal and designed a printable tiny thermocouple base on them. This thermocouple showed an excellent linear property when the temperature ranged from 0 to 200 °C [109]. Likewise, some force sensors were fabricated on the basis of resistive response, including pressure sensors [95,110], strain sensors [97,111] and shear force sensors [111]. To accurately measure the change of resistance, several specific circuit organizations were applied. Figure 7B is the schematic of a Wheatstone bridge with a Galinstan based sensor. The application of a Wheatstone bridge helps detect resistance of the sensor with high precision. When the voltage V_a is equal to V_b, which means that the output sensor V_{out} is zero, the resistance of the Galinstan based sensor can be calculated by the following equation:

$$R_x = \frac{R_2}{R1}R_0\tag{3}$$

V_{out} is the amplification of $V_b - V_a$. This circuit can sensitively reflect the change of voltage and thus the resistance of the sensor. For instance, when applying external force to the Galinstan-based pressure sensor, deformation of the soft circuit would cause the change of resistance, which further influences the output voltage V_{out}. The resistance of R_0 should be adjusted until V_{out} is zero again. Then the new resistance of the pressure sensor can be calculated, as well as the external force.

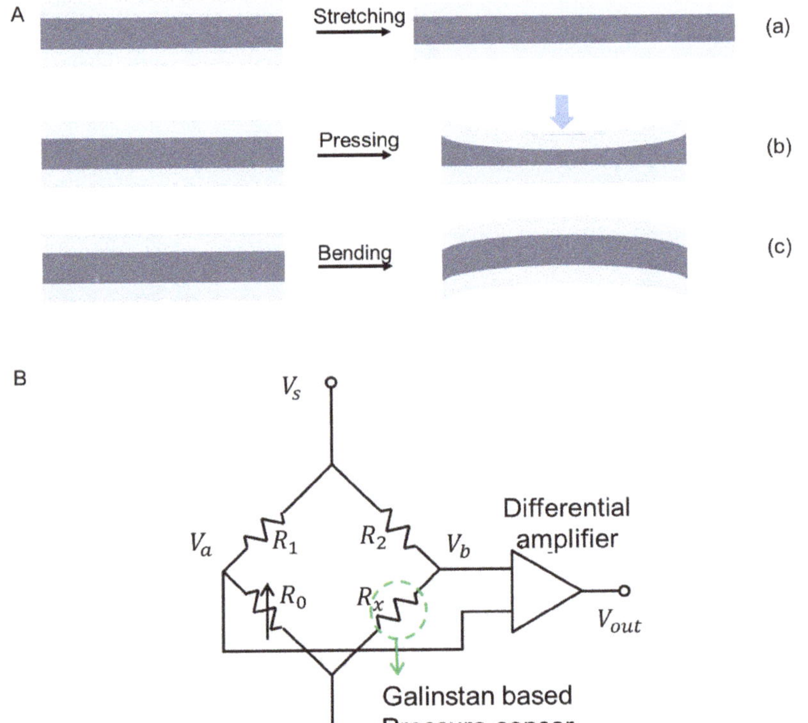

Figure 7. Working principle of resistive sensor. (**A**) Deformation of liquid metal under external force. Blue represents soft substrate and gray represents liquid metal: (**a**) Stretching; (**b**) Pressing; (**c**) Bending. (**B**) Wheatstone bridge to detect the resistance of the liquid-metal pressure sensor, reproduced with permission from [95].

4.3. Capacitive Sensors

Some capacitive force sensors have been designed as well, which mainly work based on the variation of capacitance [66,112–115]. The theory of capacitive sensors is similar to that of resistive sensors. In principle, the capacitance is decided by the formula below:

$$C = (\varepsilon S)/d, \tag{4}$$

where ε is the dielectric constant; S is the facing area and d is the vertical distance (Figure 8A). When applying an external force, parameters of the capacitor will be altered, causing the change of capacitance.

More complex and precise sensors can be fabricated through the series and parallel combination of capacitors. Figure 8B is a stretchable sensor with two liquid metal fibers, whose capacitance would change when being twisted. It could be used to measure torsion, strain and touch through the change of capacitance. Moreover, an inertial sensor based on a liquid metal droplet was designed as well. The liquid droplet was not fixed, so that it could modulate the capacitance between electrode 1 and 2 during movement. Compared to rigid sensors, these capacitive sensors based on liquid metal has good flexibility and stability.

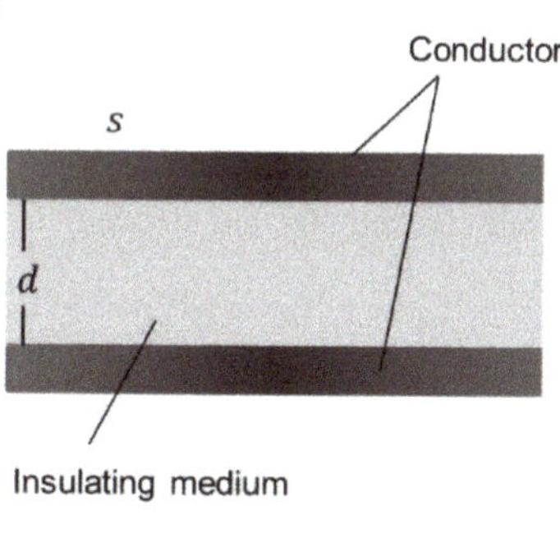

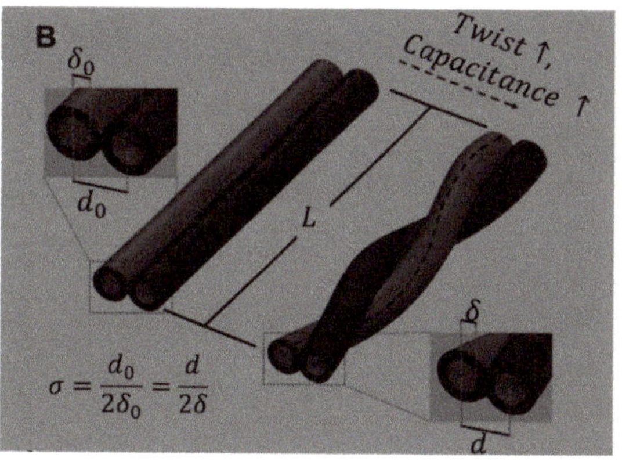

$$\sigma = \frac{d_0}{2\delta_0} = \frac{d}{2\delta}$$

Figure 8. Working principle of capacitive sensor. (**A**) Illustration of capacitor: s is the facing area and d is the vertical distance. (**B**) Stretchable capacitive sensors with double helix liquid metal fibers, reproduced with permission from [114].

4.4. Electrochemical Sensors

Electrochemical sensors utilize the chemical reaction occurring in solution. EGaIn shows good electrochemical behavior. Liquid metal with nickel and aluminum has been proven to be propelled by hydrogen in NaOH solution [78,116]. The movement can be controlled by an external electrical or magnetic field. Combined with saline solution, liquid metal can oxidize quickly and become gray gradually. Based on these reactions, liquid metal can be used to detect the existence of some substance and electrical field.

Moreover, Sivan et al. found that liquid-metal marbles could enhance the sensitivity of heavy metal ion sensors [117]. They produced liquid-metal marbles with controlled coating density and proved that these marbles showed enhanced electrochemical sensitivity towards heavy metal ions with excellent selectivity. It had been used to detect some specific heavy metal ions, such as Pb^{2+} or Cd^{2+}. Compared to resistive sensors and capacitive sensors, studies on electrochemical sensors are relatively fewer.

4.5. Metamaterial Biosensors

With the development of metamaterials, liquid-metal metamaterials are on the rise. Among all the metamaterials, split-ring resonator (SRR) arrays and complementary split-ring resonator arrays (CSRR) are the most commonly used structures for electromagnetic metamaterials, as shown in Figure 9A,B. Single SRR unit and SRR arrays based on liquid-metal have been fabricated successfully by printing (Figure 9C,D). All those results prove that liquid-metal based metamaterials are possible. The primary limitation of a liquid-metal SRR structure is that the fabrication size is not fine enough to generate electromagnetic response. Besides SRR and CSRR structures, several other types of liquid metal electromagnetic metamaterials have been proposed as well. For instance, Xu et al. fabricated a switchable metamaterial exploiting liquid metal [118]. The interference between copper and liquid metal resulted in an electromagnetically induced transparency (EIT)-like spectrum, which could be switched on or off on the basis of the fluidity of liquid metal. Moreover, a flexible metamaterial absorber based on liquid metal and PDMS was proposed [119–121]. Reichel et al. also designed a terahertz signal-processing device with liquid metal, which was electrically reconfigurable [122].

Recent studies have also applied metamaterials to biosensors for molecule detection. The sensing mechanism of metamaterials is that a change in the resonant frequency would happen when different biomolecules bind onto metamaterial resonators [123]. It is hoped that metamaterials can be combined with with liquid metal to design soft metamaterial biosensors. Generally, research on electromagnetic metamaterials based on liquid metal are still in the initial stage and more efforts are needed. Upcoming research could be conducted to find out the feasibility of liquid-metal metamaterial biosensors.

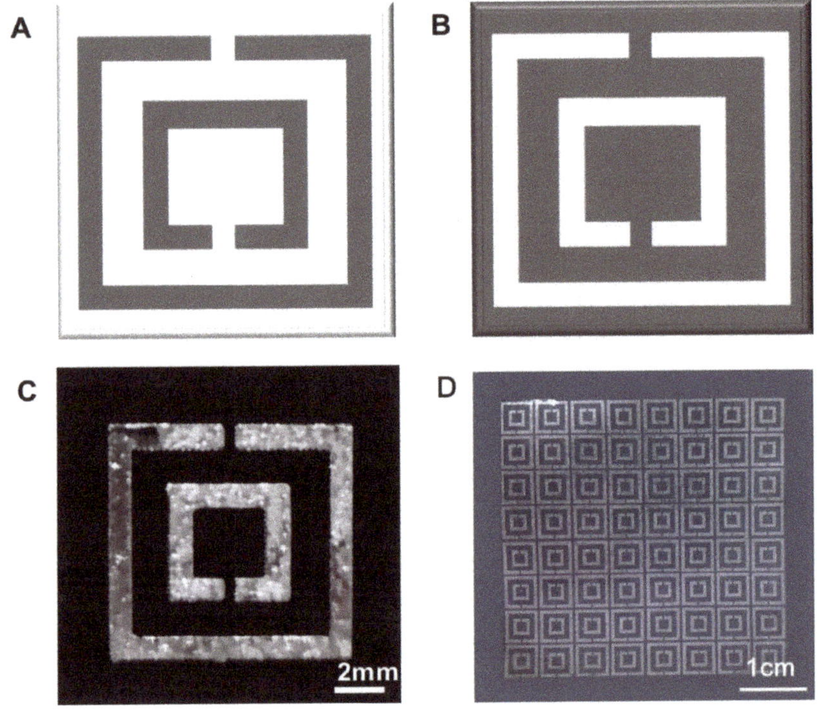

Figure 9. Metamaterial. Schematic diagram of split-ring resonator (SRR) (**A**) and complementary split-ring resonator (CSRR) (**B**), where the gray color represents metal and the white represents gaps. (**C**) An SRR unit fabricated by liquid metal. (**D**) SRR arrays fabricated by liquid metal.

4.6. Liquid-Metal Antenna

Antennas are important components in wireless devices for remote communication and sensing [124]. Several flexible antennas fabricated by liquid metal have been designed for implanted medical devices, interactive gaming and aeronautic remote sensing to avoid the mechanical failure of solid parts due to material fatigue, creep or wear [124–127]. A radiation pattern reconfigurable antenna was designed by Rodrigo et al., which could operate at 1800 MHz with 4.0% bandwidth [124]. Mazlouman et al. produced a frequency-reconfigurable antenna as well by injecting liquid metal into a square reservoir. Their antenna allowed frequency tuning from 1.3 to 3 GHz. Hayes et al. proposed a flexible antenna with a novel multi-layer construction based on liquid metal [126]. It was proven to be mechanically flexible and perform stably during flexing. Cheng et al. fabricated a flexible planar inverted cone antenna for ultrawideband applications [127]. It is expected that antennas based on liquid metal with improved durability and stability will be presented in the near future.

5. Typical Applications

Sensors play an important role in biomedicine and bionic area. So far, the application of soft sensors includes force sensors, temperature sensors, blood glucose sensors and so on. Based on the fluidity and metallicity of liquid metal, sensors enabled from them possess good flexibility and sensitivity. Obviously, sensors fabricated by liquid metal will become critical components for wearable devices and robots.

5.1. Force Sensors

Soft force sensors based on liquid metal, which can tolerate large deformation, have shown advantages in force detection and motion-monitoring.

For instance, pressure sensors have been researched vigorously. Previous experiments showed that when directly applying pressure to a printed liquid metal pressure sensor, the resistance would change accordingly [110]. As shown in Figure 10A, when there existed external pressure, the resistance of the liquid-metal sensor would increase. Moreover, Jung et al. manufactured a pressure sensor embedded in a microfluidic channel, which could be used to measure the viscosity of various fluids [95]. The change of the fluids' viscosity would lead to a pressure to the membrane and then the electrical resistance would increase due to the decrease in the cross-sectional area (Figure 10B). A Wheatstone bridge circuit was applied in their research to detect the change of electrical resistance accurately. Kim et al. integrated liquid metal microchannels with a 3D-printed microbump array to increase the sensitivity of the pressure sensor, which has been proven to have a wide range of applications in health monitoring [93]. Oh et al. injected liquid metal into PDMS and designed a pressure-conductive rubber sensor. This could be used to detect wrist pulse and respiration [128]. Besides pressure sensors, strain sensors are also important types of force sensors. For instance, Zhou et al. fabricated a flexible strain sensor with asymmetric structure. This could detect various human joint motions simultaneously [129].

Vogt et al. have designed a multi-axis force sensor with microfluidic channels recently [106]. This could detect not only normal shear force, but also in-plane shear force. Meanwhile, Shi et al. reported a flexible force sensor to detect normal and shear forces as well based on the piezoresistive effect and further designed an artificial hair cell sensor [130]. Figure 10C shows the working principle of the hair cell sensor. The polymer hair would be deformed when there was a shear or normal force, causing the deformation of circuits connected to the hair. The change of capacitance between liquid metal could reflect the movement of polymer hair. It perfectly imitated the operating mode of hair cells in the inner ear, which might further propel the research of an artificial ear.

Cooper et al. proposed a stretchable capacitive sensor with double-helix liquid-metal fibers, which was capable of detecting torsion, strain and touch [114]. When being twisted or elongated, the geometry of fibers as well as the capacitance between the fibers would change accordingly. Moreover, a soft inertial sensor was introduced by Varga et al. based on the alteration of capacitance between two electrodes when the liquid-metal droplet moved [66]. This inertial sensor was sensitive to motion and position detection.

Recently, wearable transient epidermal sensors were fabricated by liquid-metal hydrogel [131]. Liquid-metal marbles were distributed evenly into polyvinyl alcohol by sonication to form liquid metal hydrogel. This material was flexible and self-healing. Epidermal sensors made by it could detect some human activities, like swallowing and finger bending.

Furthermore, soft force sensors are expected to be applied to electronic skin and intelligent robots to further promote the function of artificial products. Combining the advantages of traditional artificial skin and sensors, a soft electronic skin can provide not only aesthetic function, but also a perception function [132,133].

5.2. Temperature Sensors

Several types of temperature sensors based on liquid metal have been studied. Hong et al. applied liquid metal as the interconnection between temperature-sensing units to fabricate soft temperature sensor arrays [106]. Their sensor array could be easily attached to the skin, which was important for electronic skin as well. A thermocouple based on liquid metal was proposed by Li et al. [109]. Ga with 0.25 wt. % Ga oxides and EGaIn21.5 with 0.25 wt. % Ga oxides were chosen as the thermocouple elements. The thermocouple was directly printed on paper (Figure 10D) and could be bent and twisted to some extent.

Currently, temperature sensors based on liquid metal have not been investigated deeply because compared to existing rigid metal temperature sensors, liquid metal temperature sensors are not sensitive and convenient enough. Tremendous efforts will be made in the near future.

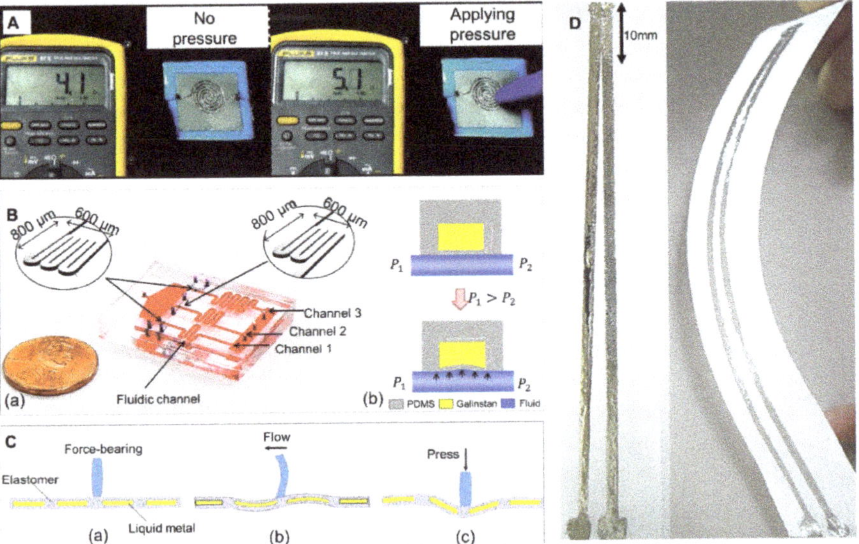

Figure 10. Photos of force sensors, temperature sensors and blood glucose sensors. (**A**) The resistance of a printed pressure sensor according to external pressure, reproduced with permission from [110]. (**B**) A micro-channel pressure sensor: (**a**) Structure of the sensor; (**b**) Working principle of measuring viscosity of various fluids; Reproduced with permission from [95]. (**C**) Schematic of artificial hair cell sensor: (**a**) No external force; (**b**) Deformation under a shear force; (**c**) Deformation under a normal force. (**D**) Thermocouple based on liquid metal printed on paper, reproduced with permission from [109].

5.3. Blood Glucose Sensors

Yi et al. have designed an electrochemical sensor for wireless glucose-detecting based on liquid metal [134]. They fabricated liquid-metal electrodes by direct hand printing on polyvinyl chloride (PVC) substrate. In their research, those electrodes were modified by enzyme to detect glucose. When the enzyme reacted with glucose, the electrodes could transfer it to an electrical signal. Overall, this flexible sensor has been proven to be reliable by cycle voltammetry. They also designed a wireless system to transmit testing data to a smart phone via Bluetooth.

5.4. Sensor Array

An array is a common circuit arrangement for sensors, which can improve their sensitivity and accuracy. For instance, the temperature sensors [106] and the pressure sensor [95] mentioned above are all in the form of arrays. Specially, Zhang et al. developed a sensor array for application in robotic electronic skin [135]. They used cross-grid liquid-metal layers as electrodes. Both capacitance mode and triboelectric nanogenerators (TENG) [136] mode could be realized by this sensor array. There were two layers of cross-grid arrays constructed by liquid-metal electrodes. The capacitance between the top and bottom electrodes would increase or decrease when the environment changed. The adoption of liquid metal made the whole device highly flexible. It could function as an E-skin sensor with complementary capacitance mode or TENG mode, respectively. Compared to a single sensing element, sensor arrays can provide large-area and multi-functional detecting, especially for human health-monitoring.

5.5. Pneumatic Artificial Muscles

Pneumatic artificial muscles have been increasingly investigated owing to the potential in wearable devices. Inspired by biological muscle, Jackson et al. designed an artificial muscle with both force and position sensors [137], as shown in Figure 11A. The Golgi tendon organ and muscle spindle were fabricated by EGaIn. Figure 11B shows the form alterations when the sensorized, flat, pneumatic artificial muscle was at rest and inflated to the pressure of 82.8 kPa. Liquid metal was injected into micro-channels to form contractile and force sensors (Figure 11C). It has been proved that this artificial muscle could mimic the behavior of biological muscle well. The progress in artificial muscle is meaningful to wearable devices, as well as soft robots.

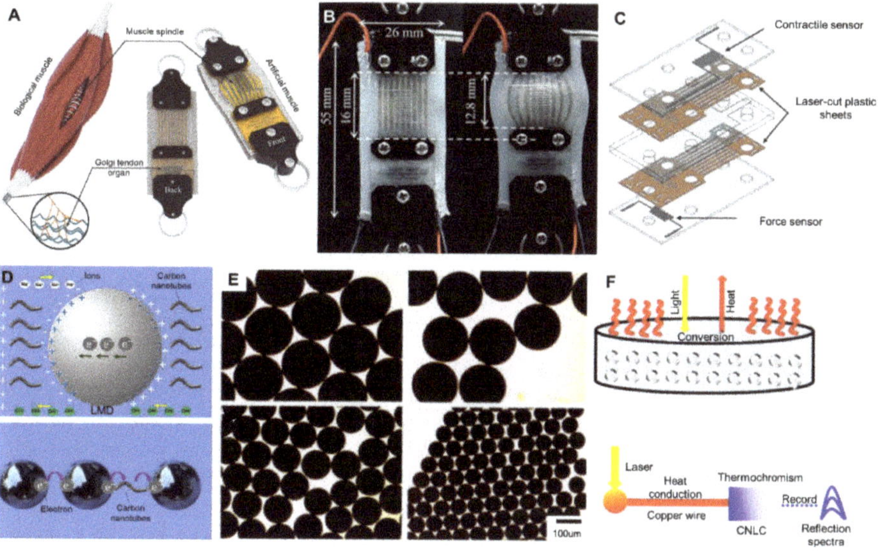

Figure 11. Artificial muscles and liquid metal microsphere sensors. (**A**) Schematic comparison between biological muscle and artificial muscle, reproduced with permission from [136]. (**B**) The sensorized, flat, pneumatic artificial muscle at rest and inflated to 82.8 kPa, reproduced with permission from [137]. (**C**) Illustration of the contraction and pressure sensors, reproduced with permission from [137]. (**D**) The electrical environment of a liquid-metal microsphere in NaOH solution (above) and two methods of communicating for liquid-metal microspheres in a circuit (below), reproduced with permission from [77]. (**E**) Microscopy images showing liquid-metal microspheres produced by capillary-based microfluidics, reproduced with permission from [77]. (**F**) Schematic diagram of the photothermal conversion of liquid-metal microspheres and the near-infrared (NIR) sensor based on the liquid metal microsphere, reproduced with permission from [138].

5.6. Liquid-Metal Microsphere Sensors

The potential application of liquid-metal microspheres is increasingly attractive as well. The concept of liquid-metal enabled droplet circuits in solution was proposed previously [77]. Liquid metal droplets, ions and carbon nanotubes were combined in electrolyte solution to enhance electrical conductivity. Figure 11D shows the electric charge distribution of a liquid-metal microsphere in NaOH solution, and the communication methods between liquid-metal microspheres and carbon nanotubes in a circuit. The quantum tunneling effect is the fundamental principle lying behind liquid-metal droplet circuits. Besides the theoretical liquid-metal droplet circuits, sensors based on liquid-metal microsphere were researched recently [138]. Wei et al. designed a method to fabricate liquid-metal microspheres with different sizes through a capillary effect, as shown in Figure 11E. The size of liquid-metal microspheres could be precisely and stably controlled by the co-flowing configuration in the micro-channels. These liquid-metal microspheres had been proven to exhibit outstanding photothermal properties, which could be utilized to produce optical sensors, such as near infrared ray (NIR) sensors (Figure 11F). There is a photo-thermal conversion in liquid metal microsphere, which is promising to not only liquid-metal optical sensors, but also light-driven actuators and as an energy conversion medium. Moreover, Zhang et al. presented a flexible capacitive sensor based on liquid-metal microsphere as well [138]. It worked through the multi-plateau capacitance waveform when liquid metal sphere passed through the sensing area. It is believed that more applications of liquid-metal microspheres could be explored.

6. Perspective

In the near future, we may expect the appearance of metal-enabled liquid sensors. By contrast with traditional rigid and soft sensors, metal-enabled liquid sensors will be in the form of fluidity on the basis of liquid metal. The environment of the human body is liquid as well and all organs work in this environment. According to the ideas of bionics, research on liquid sensors may be an inevitable trend.

6.1. Concept of Liquid Sensors

As shown in Figure 12A, a liquid sensor is fabricated by wrapping liquid metal with a flexible membrane. Liquid metal inside the membrane has greater freedom than that printed on soft substrate or injected into micro-channels. Thus, the liquid sensors are more flexible and transformable. To ensure good mechanical property of liquid sensors and avoid leakage of liquid metal, the membrane must exhibit excellent flexibility and compactness. Figure 12B gives the details of a liquid-sensor system. Processing circuits are required in order to amplify, transform and filtrate the electrical signal output by the liquid sensors. Then, the processed signal could be measured by the detecting device. Compared to existing soft sensors, liquid sensors are more suitable for in vivo detection because it is highly transformable. Injectable electrodes based on liquid metal have been proposed for tumor treatment already [65]. It is possible to design injectable liquid sensors with the application of liquid metal.

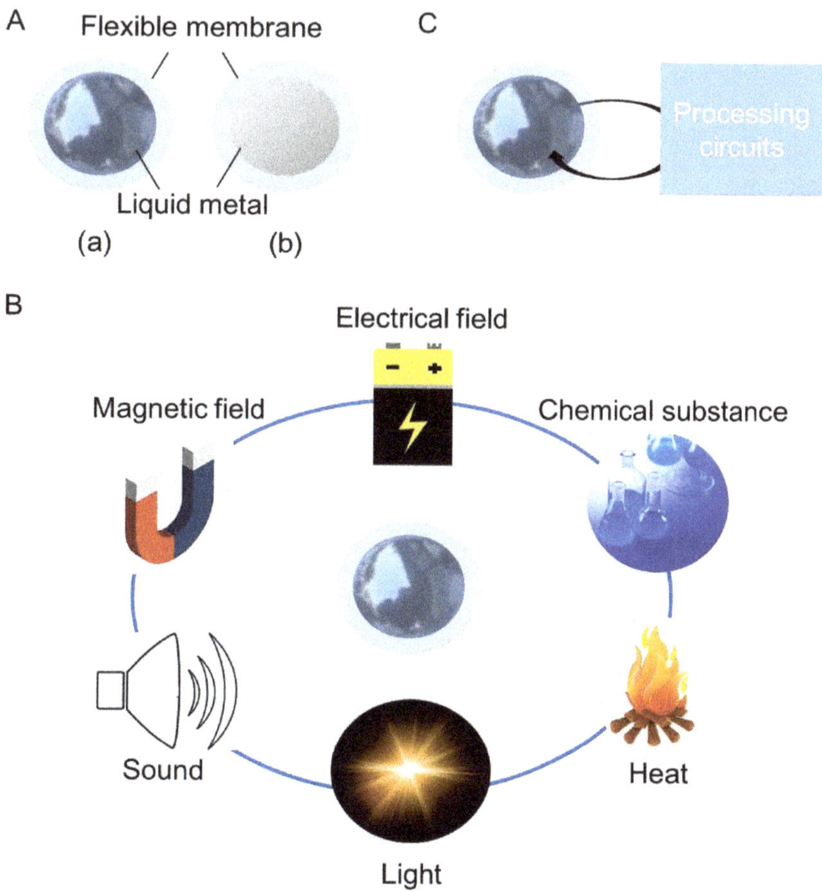

Figure 12. Metal-enabled liquid sensors. (**A**) Illustration of liquid sensor: (**a**) Front view; (**b**) Sectional view. (**B**) Working principle of liquid sensor. (**C**) Diagram of a liquid sensor system.

6.2. Working Principle of Liquid Sensors

Metal-enabled liquid sensors can make a response to external force, electrical field, magnetic field, sound, light, heat and some chemical substances, on the basis of the unique physical and chemical properties of liquid metal (Figure 12C). External stimulations will result in the changes of resistance or capacitance, which can be converted to measurable electrical signals, such as voltage or current signal. To reflect the alteration of the environment, the liquid sensors should be connected to a processing unit, as shown in Figure 12C.

Qualitative liquid sensors based on liquid metal are not hard to realize. The difficulty is how to create a quantitative liquid sensor with high sensitivity and precision. More efforts are required to propel the development of liquid sensors.

7. Discussion

Liquid metal is characterized by good fluidity, perfect metallicity and unique electromagnetic properties. Liquid metal-based soft sensors have attracted increasing attention. Generally, soft sensors fabricated by liquid metal imply the following advantages.

Firstly, sensors fabricated by liquid metal own excellent flexibility owing to the fluidity of liquid metal. They can remain working when being twisted, stretched and bent within limits. The existence of liquid metal ensures that the circuits are still connected during deformation and improves the reliability of those soft sensors. This allows the application of liquid-metal sensors in wearable devices with great promise.

Secondly, fabrication methods of liquid-metal based sensors are diverse. Due to the unique physical and chemical properties of liquid metal, patterning methods includes printing, microfluidic technology and others. Compared to traditional micro-/nano-machining technology, these new methods are convenient and economical. However, the resolution should be further improved by optimizing the fabrication process.

Last but not least, liquid metal-based sensors provide better conformability and sensing sensitivity. Since the flexibility of liquid metal circuit is guaranteed, it can well fit different parts of human body and promote the wearing experience. Furthermore, since sensors based on liquid metal can be tightly attached to objects to be tested, the sensing sensitivity is higher than that of rigid sensors or common soft sensors.

So far, liquid metal-based sensors have been vigorously researched because of their outstanding performance. It is hoped that more soft sensors based on liquid metal, especially metamaterial biosensors, can be designed. Through increasing endeavors, more sensing mechanisms enabled from liquid metals are also possible in the near future. Microsphere sensors based on liquid metal, and even metal-enabled liquid sensors, may cause a revolution in the development of soft sensors and in vivo human signal monitoring.

8. Conclusion

This article provides a systematic review on liquid metal-enabled soft sensors as well as possible future development. Typical sensors made in this way include force sensors, temperature sensors, electrochemical sensors and so on. They work mainly depending on the alteration of resistance and capacitance of liquid metal. Electromagnetic metamaterials fabricated by liquid metal may propel the development of metamaterials and flexible metamaterial biosensors. With the excellent fluidity of liquid metal, those sensors show excellent stretchability, which have potential in wearable device and health-monitoring. Electrode arrays can be combined with soft sensors for large-area detection and sensitivity improvement.

In summary, liquid metal has shown great potential in the area of soft sensors. Based on the progress achieved, more efforts are necessary and the efforts will be worthy. Future research may focus on resolution and repeatability as well as designing the testing principles of liquid metal-based sensors. Microsphere sensors based on liquid metal and metal-enabled liquid sensors will be important research fields in the future. It is believed that the soft sensors fabricated by liquid metal could be widely applied in diverse biomedical areas and soft robots.

Author Contributions: Conceptualization, Y.R. and J.L.; literature review, Y.R.; writing—original draft preparation, Y.R.; writing—review and editing, Y.R., X.S. and J.L.; visualization, Y.R.; supervision, J.L.; project administration, J.L.; funding acquisition, J.L. All authors have read and agreed to the published version of the manuscript.

Funding: This work was funded by the NSFC Key Project under Grant No. 51890893, No.91748206, Dean's Research Funding and the Frontier Project of the Chinese Academy of Science.

Conflicts of Interest: The authors declare no conflict of interest.

References

1. Webster, J.G. *Medical Instrumentation: Application and Design*; John Wiley & Sons Inc.: Hoboken, NJ, USA, 1998.
2. Harsanyi, G. *Sensors in Biomedical Applications: Fundamentals, Technology and Applications*; CRC Press: Boca Raton, FL, USA, 2000.

3. Gong, S.; Schwalb, W.; Wang, Y.; Chen, Y.; Tang, Y.; Si, J.; Shirinzadeh, B.; Cheng, W. A wearable and highly sensitive pressure sensor with ultrathin gold nanowires. *Nat. Commun.* **2014**, *5*, 1–8. [CrossRef] [PubMed]
4. Viry, L.; Levi, A.; Totaro, M.; Mondini, A.; Mattoli, V.; Mazzolai, B. Flexible three-axial force sensor for soft and highly sensitive artificial touch. *Adv. Mater.* **2014**, *26*, 2659–2664. [CrossRef] [PubMed]
5. Water, W.; Chen, S.E. Using ZnO nanorods to enhance sensitivity of liquid sensor. *Sensor. Actuat. B Chem.* **2009**, *136*, 371–375. [CrossRef]
6. Schneditz, D.; Kenner, T.; Heimel, H.; Stabinger, H. A sound-speed sensor for the measurement of total protein concentration in disposable, blood-perfused tubes. *J. Acoust. Soc. Am.* **1989**, *86*, 2073. [CrossRef]
7. Zhang, H.; Yang, Y.; Su, Y.; Chen, J.; Adams, K.; Lee, S.; Hu, C.; Wang, Z. Triboelectric nanogenerator for harvesting vibration energy in full space and as self-powered acceleration sensor. *Adv. Funct. Mater.* **2014**, *24*, 1401–1407. [CrossRef]
8. Peet, D.J.; Pryor, M.D. Evaluation of a mosfet radiation sensor for the measurement of entrance surface dose in diagnostic radiology. *Br. J. Radiol.* **1999**, *72*, 562–568. [CrossRef]
9. Yang, H.; Qi, D.; Liu, Z.; Chandran, B.K.; Wang, T.; Yu, J.; Chen, X. Soft thermal sensor with mechanical adaptability. *Adv. Mater.* **2016**, *28*, 9175–9181. [CrossRef]
10. Wang, C.F.; Wang, C.H.; Huang, Z.L.; Xu, S. Materials and structures toward soft electronics. *Adv. Mater.* **2018**, *30*. [CrossRef]
11. Koo, J.H.; Song, J.; Kim, D. Solution-processed thin films of semiconducting carbon nanotubes and their application to soft electronics. *Nanotechnology* **2019**, *30*, 13. [CrossRef]
12. Djenizian, T.; Tee, B.; Ramuz, M.; Fang, L. Advances in flexible and soft electronics. *APL Mater.* **2019**, *7*. [CrossRef]
13. Kang, B.; Lee, S.K.; Jung, J.; Joe, M.; Lee, S.B.; Kim, J.; Lee, C.; Cho, K. Nanopatched graphene with molecular self-assembly toward graphene-organic hybrid soft electronics. *Adv. Mater.* **2018**, *30*. [CrossRef] [PubMed]
14. Zhang, S.; Wang, B.; Jiang, J.; Wu, K.; Guo, C.F.; Wu, Z. High fidelity conformal printing of 3D liquid alloy circuits for soft electronics. *ACS Appl. Mater. Inter.* **2019**, *11*, 7148–7156. [CrossRef] [PubMed]
15. Valentine, A.D.; Busbee, T.A.; Boley, J.W.; Raney, J.R.; Lewis, J.A. Hybrid 3D printing of soft electronics. *Adv. Mater.* **2017**, *29*, 29. [CrossRef] [PubMed]
16. Wang, C.; Xia, K.; Zhang, Y.; Kaplan, D.L. Silk-based advanced materials for soft electronics. *Acc. Chem. Res.* **2019**, *52*, 2916–2927. [CrossRef]
17. Feng, Y.; Zhu, J. Copper nanomaterials and assemblies for soft electronics. *Sci. China Mater.* **2019**, *62*, 1679–1708. [CrossRef]
18. Kang, J.; Tok, J.; Bao, Z. Self-healing soft electronics. *Nat. Electron.* **2019**, *2*, 144–150. [CrossRef]
19. Li, R.; Wang, L.; Yin, L. Materials and devices for biodegradable and soft biomedical electronics. *Materials* **2018**, *11*, 2108. [CrossRef]
20. Choi, C.; Choi, M.K.; Hyeon, T.; Kim, D. Nanomaterial-based soft electronics for healthcare applications. *ChemNanoMat* **2016**, *2*. [CrossRef]
21. Hoshi, T. Robot skin based on touch-area-sensitive tactile element. In Proceedings of the Presented at IEEE International Conference on Robotics & Automation, Orlando, FL, USA, 15–19 May 2006.
22. Bartlett, M.D.; Markvicka, E.J.; Majidi, C. Rapid fabrication of soft, multilayered electronics for wearable biomonitoring. *Adv. Funct. Mater.* **2016**, *26*, 8496–8504. [CrossRef]
23. Gong, S.; Lai, D.; Su, B.; Si, K.J.; Ma, Z.; Yap, L.W.; Guo, P.; Cheng, W. Highly stretchy black gold E-skin nanopatches as highly sensitive wearable biomedical sensors. *Adv. Electron. Mater.* **2015**, *1*. [CrossRef]
24. Jeong, S.H.; Zhang, S.; Hjort, K.; Hilborn, J.; Wu, Z. PDMS-based elastomer tuned soft, stretchable, and sticky for epidermal electronics. *Adv. Mater.* **2016**, *28*, 5830–5836. [CrossRef] [PubMed]
25. Yuan, W.; Dong, S.; Adelson, E. GelSight: High-resolution robot tactile sensors for estimating geometry and force. *Sensors* **2017**, *17*, 2762. [CrossRef] [PubMed]
26. Tian, L.; Li, Y.; Webb, R.; Krishnan, S.; Bian, Z.; Song, J.; Ning, X.; Crawford, K.; Kurniawan, J.F.; Bofinas, A.P.; et al. Flexible and stretchable 3ω sensors for thermal characterization of human skin. *Adv. Funct. Mater.* **2017**, *27*, 26. [CrossRef]
27. Sonar, H.A.; Paik, J. Soft pneumatic actuator skin with piezoelectric sensors for vibrotactile feedback. *Front. Robot. AI* **2016**, *2*, 10–3389. [CrossRef]

28. Hu, T.; Xuan, S.; Ding, L.; Gong, X. Stretchable and magneto-sensitive strain sensor based on silver nanowire-polyurethane sponge enhanced magnetorheological elastomer. *Mater. Des.* **2018**, *156*, 528–537. [CrossRef]

29. Jiong, Y.; Cheng, W.; Kourosh, K. Electronic skins based on liquid metals. *Proc. IEEE* **2019**, *107*, 2168–2184.

30. Stoppa, M.; Chiolerio, A. Wearable electronics and smart textiles: A critical review. *Sensors* **2014**, *14*, 11957–11992. [CrossRef]

31. Rajan, K.; Garofalo, E.; Chiolerio, A. Wearable Intrinsically Soft, Stretchable, Flexible Devices for Memories and Computing. *Sensors* **2018**, *18*, 367. [CrossRef]

32. Roh, E.; Hwang, B.U.; Kim, D.; Kim, B.Y.; Lee, N.E. Stretchable, transparent, ultrasensitive, and patchable strain sensor for human–machine interfaces comprising a nanohybrid of carbon nanotubes and conductive elastomers. *ACS Nano* **2015**, *9*, 6252–6261. [CrossRef]

33. Chen, S.; Jiang, K.; Lou, Z.; Chen, D.; Shen, G. Recent developments in graphene-based tactile sensors and e-skins. *Adv. Mater. Technol.* **2017**, *3*. [CrossRef]

34. Wang, C.; Xia, K.; Zhang, M.; Jian, M.; Zhang, Y. An all silk-derived, dual-mode e-skin for simultaneous temperature-pressure detection. *ACS Appl. Mater. Inter.* **2017**, *9*, 39484–39492. [CrossRef] [PubMed]

35. Tajima, R.; Kagami, S.; Inaba, M.; Inoue, H. Development of soft and distributed tactile sensors and the application to a humanoid robot. *Adv. Robot.* **2002**, *16*, 381–397. [CrossRef]

36. Ozel, S.; Keskin, N.A.; Khea, D.; Onal, C.D. A precise embedded curvature sensor module for soft-bodied robots. *Sensor. Actuat. A Phys.* **2015**, *236*, 349–356. [CrossRef]

37. Li, S.; Zhao, H.; Shepherd, R.F. Flexible and stretchable sensors for fluidic elastomer actuated soft robots. *MRS Bull.* **2017**, *42*, 138–142. [CrossRef]

38. Yeo, J.C.; Lim, C.T. Emerging flexible and wearable physical sensing platforms for healthcare and biomedical applications. *Microsyst. Nanoeng.* **2016**, *2*. [CrossRef]

39. Dinh, T.; Phan, H.; Nguyen, K.; Qamar, A.; Md Foisal, A.R.; Viet, T.; Tran, C.D.; Zhu, Y.; Nguyen, N.T.; Dao, D.V. Environment-friendly carbon nanotube based flexible electronics for noninvasive and wearable healthcare. *J. Mater. Chem. C* **2016**, *4*. [CrossRef]

40. Wang, D.; Mei, Y.; Huang, G. Printable inorganic nanomaterials for flexible transparent electrodes: From synthesis to application. *J. Semicond.* **2018**, *39*. [CrossRef]

41. Fu, K.; Yao, Y.; Dai, J.; Hu, L. Progress in 3D printing of carbon materials for energy-related applications. *Adv. Mater.* **2016**, *29*. [CrossRef]

42. Choi, Y.; Kang, J.; Secor, E.; Sun, J.; Kim, H.; Lim, J. Capacitively coupled hybrid ion gel and carbon nanotube thin-film transistors for low voltage flexible logic circuits. *Adv. Funct. Mater.* **2018**, *28*, 28. [CrossRef]

43. Zhang, D.; Chi, B.; Li, B.; Gao, Z.; Du, Y.; Guo, J.; Wei, J. Fabrication of highly conductive graphene flexible circuits by 3D printing. *Synth. Met.* **2016**, *27*, 79–86. [CrossRef]

44. Saidina, D.; Eawwiboonthanakit, N.; Jaafar, M.; Fontana, S.; Hérold, C. Recent development of graphene-based ink and other conductive material-based inks for flexible electronics. *J. Electron. Mater.* **2019**, *48*, 3428–3450. [CrossRef]

45. Yang, W.; Wang, C. Graphene and the related conductive inks for flexible electronics. *J. Mater. Chem. C* **2016**, *4*, 7193–7207. [CrossRef]

46. Park, J.; Zhou, C.; Yang, C.Y. Carbon-based Nanostructures for Flexible Electronics. In Proceedings of the IEEE International Conference on Electron Devices and Solid State Circuits, Shenzhen, China, 6–8 June 2018.

47. Hao, Y.; Gao, J.; Xu, Z.; Zhang, N.; Luo, J.; Liu, X. Preparation of silver nanoparticles with hyperbranched polymers as a stabilizer for inkjet printing of flexible circuits. *New J. Chem.* **2019**, *43*, 2797–2803. [CrossRef]

48. Stephan, T.D.; Vimolvan, P. Humic acid assisted synthesis of silver nanoparticles and its application to herbicide detection. *Mater. Lett.* **2008**, *62*, 2661–2663.

49. Sun, J.; Zhou, W.; Yang, H.; Zhen, X.; Ma, L.; Williams, D.; Sun, D.; Lang, M. Highly transparent and flexible circuits through patterning silver nanowires into microfluidic channels. *Chem. Commun.* **2018**, *54*, 4923–4926. [CrossRef]

50. Seung, H.K. Low temperature thermal engineering of nanoparticle ink for flexible electronics applications. *Semicond. Sci. Technol.* **2016**, *31*, 31. [CrossRef]

51. Tan, S.; Zu, X.; Yi, G.; Liu, X. Synthesis of highly environmental stable copper-silver core-shell nanoparticles for direct writing flexible electronics. *J. Mater. Sci. Mater. Electron.* **2017**, *28*, 15899–15906. [CrossRef]

52. Shao, W.; Li, G.; Zhu, P.; Zhang, Y.; Ouyang, Q.; Sun, R.; Chen, C.; Wong, C.P. Facile synthesis of low temperature sintering Ag nanopaticles for printed flexible electronics. *J. Mater. Sci. Mater. Electron.* **2018**, *29*, 4432–4440. [CrossRef]

53. Chiolerio, A.; Rajan Roppolo, I.; Chiappone, A.; Bocchini, S.; Perrone, D. Silver nanoparticle ink technology: State of the art. *Nanotech. Sci. Appl.* **2016**, *9*, 1. [CrossRef]

54. Yi, L.; Liu, J. Liquid metal biomaterials: A newly emerging area to tackle modern biomedical challenges. *Inter. Mater. Rev.* **2017**, *62*, 1–26. [CrossRef]

55. Wang, Q.; Yu, Y.; Liu, J. Preparations, characteristics and applications of the functional liquid metal materials. *Adv. Eng. Mater.* **2017**, *20*. [CrossRef]

56. Jackson, N.; Buckley, J.; Clarke, C.; Stam, F. Manufacturing methods of stretchable liquid metal-based antenna. *Microsyst. Technol.* **2019**, *25*, 3175–3184. [CrossRef]

57. Ordonez, R.; Hayashi, C.; Torres, C.; Hafner, N.; Adleman, J.; Acosta, N.; Melcher, J.; Kamin, N.M.; Garmire, D. Conformal Liquid-Metal Electrodes for Flexible Graphene Device Interconnects. *IEEE Trans. Electron Dev.* **2016**, *63*, 4018–4023. [CrossRef]

58. Li, Q.; Lin, J.; Liu, T.; Zheng, H.; Liu, J. Printed flexible thin-film transistors based on different types of modified liquid metal with good mobility. *Sci. China Informa. Sci.* **2019**, *62*. [CrossRef]

59. Nayak, S.; Li, Y.; Tay, W.; Zamburg, E.; Singh, D.; Lee, C.; Koh, S.J.A.; Chia, P.; Thean, A.V.Y. Liquid-metal-elastomer foam for moldable multi-functional triboelectric energy harvesting and force sensing. *Nano Energy* **2019**, *64*. [CrossRef]

60. Kim, S.; Oh, J.; Jeong, D.; Bae, J. Direct wiring of eutectic gallium-indium to a metal electrode for soft sensor systems. *ACS Appl. Mater. Inter.* **2019**, *11*, 20557–20565. [CrossRef] [PubMed]

61. Kim, T.; Kim, D.; Lee, B.; Lee, J. Soft and Deformable Sensors Based on Liquid Metals. *Sensors* **2019**, *19*, 4250. [CrossRef]

62. Yang, Y.; Jie, Z.; Jing, L.; Sabato, D. Biomedical implementation of liquid metal ink as drawable ECG electrode and skin circuit. *PLoS ONE* **2013**, *8*. [CrossRef]

63. Zhao, Z.N.; Ju, L.; Jie, Z.; Yang, Y.; Bo, Y.; Fan, C.C.; Wang, L.; Liu, J. Liquid metal enabled flexible electronic system for eye movement tracking. *IEEE Sens. J.* **2018**, *18*, 2592–2598. [CrossRef]

64. Jin, C.; Zhang, J.; Li, X.; Yang, X.; Li, J.; Liu, J. Injectable 3-D fabrication of medical electronics at the target biological tissues. *Sci. Rep.* **2013**, *3*, 3442. [CrossRef]

65. Sun, X.; Yuan, B.; Rao, W.; Liu, J. Amorphous liquid metal electrodes enabled conformable electrochemical therapy of tumors. *Biomaterials* **2017**, *146*, 156–167. [CrossRef] [PubMed]

66. Varga, M.; Ladd, C.; Ma, S.; Holbery, J.; Tröster, G. On-skin liquid metal inertial sensor. *Lab. Chip.* **2017**, *17*, 3272–3278. [CrossRef] [PubMed]

67. Chiolerio, A.; Quadrelli, M. Smart Fluid Systems: The advent of autonomous liquid robotics. *Adv. Sci.* **2017**, *4*. [CrossRef] [PubMed]

68. Adamatzky, A.; Chiolerio, A.; Szacilowski, K. Liquid metal solves maze. *Soft Matter* 2019. [CrossRef]

69. Zhang, Q.; Zheng, Y.; Liu, J. Direct writing of electronics based on alloy and metal (dream) ink: A newly emerging area and its impact on energy, environment and health sciences. *Front. Energy.* **2012**, *6*, 311–340. [CrossRef]

70. Guo, R.; Wang, X.; Chang, H.; Yu, W.; Liang, S.; Rao, W. Ni-GaIn amalgams enabled rapid and customizable fabrication of wearable and wireless healthcare electronics. *Adv. Eng. Mater.* **2018**, *20*. [CrossRef]

71. Chang, H.; Guo, R.; Sun, Z.Q.; Wang, H.Z.; Hou, Y.; Wang, Q.; Rao, W.; Liu, J. Flexible conductive materials: Direct writing and repairable paper flexible electronics using nickel-liquid metal ink. *Adv. Mater. Interfaces* **2018**, *5*. [CrossRef]

72. Guo, R.; Yao, S.; Sun, X.; Liu, J. Semi-liquid metal and adhesion-selection enabled rolling and transfer (SMART) printing: A general method towards fast fabrication of flexible electronics. *Sci. China Mater.* **2019**, *62*, 982–994. [CrossRef]

73. Tang, J.; Zhao, X.; Li, J.; Guo, R.; Zhou, Y.; Liu, J. Gallium-based liquid metal amalgams: Transitional-state metallic mixtures (TransM2ixes) with enhanced and tunable electrical, thermal, and mechanical properties. *ACS Appl. Mater. Interfaces* **2017**, *9*, 35977–35987. [CrossRef]

74. Li, H.; Mei, S.; Wang, L.; Gao, Y.; Liu, J. Splashing phenomena of room temperature liquid metal droplet striking on the pool of the same liquid under ambient air environment. *Inter. J. Heat Fluid FL* **2014**, *47*, 1–8. [CrossRef]

75. Liu, T.; Sen, P.; Kim, C.J. Characterization of nontoxic liquid-metal alloy galinstan for applications in microdevices. *J. Microelectromech. Syst.* **2012**, *21*, 443–450. [CrossRef]

76. Kim, D.; Lee, J.B. Magnetic-field-induced liquid metal droplet manipulation. *J. Korean Phys. Soc.* **2015**, *66*, 282–286. [CrossRef]

77. Ren, Y.; Liu, J. Liquid-metal enabled droplet circuits. *Micromachines* **2018**, *9*, 218. [CrossRef] [PubMed]

78. Tan, S.C.; Yuan, B.; Liu, J. Electrical method to control the running direction and speed of self-powered tiny liquid metal motors. *Proc. Roy. Soc. A Math. Phy.* **2015**, *471*, 32–38. [CrossRef]

79. Khondoker, M.A.H.; Sameoto, D. Fabrication methods and applications of microstructured gallium based liquid metal alloys. *Smart Mater. Struct.* **2016**, *25*. [CrossRef]

80. Gao, Y.; Li, H.; Liu, J. Direct writing of flexible electronics through room temperature liquid metal ink. *PLoS ONE* **2012**, *7*. [CrossRef]

81. Wang, L.; Liu, J. Advances in the Development of Liquid Metal-Based Printed Electronic Inks. *Front. Mater.* **2019**, *6*. [CrossRef]

82. Sheng, L.; Teo, S.; Liu, J. Liquid-metal-painted stretchable capacitor sensors for wearable healthcare electronics. *J. Med. Biol. Eng.* **2016**, *36*, 265–272. [CrossRef]

83. Zheng, Y.; Zhang, Q.; Liu, J. Pervasive liquid metal based direct writing electronics with roller-ball pen. *AIP Adv.* **2013**, *3*, 6459–6463. [CrossRef]

84. Zheng, Y.; He, Z.; Gao, Y.; Liu, J. Direct desktop printed-circuits-on-paper flexible electronics. *Sci. Rep.* **2013**, *3*, 1786. [CrossRef]

85. Tabatabai, A.; Fassler, A.; Usiak, C.; Majidi, C. Liquid-phase gallium–indium alloy electronics with microcontact printing. *Langmuir* **2013**, *29*, 6194–6200. [CrossRef] [PubMed]

86. Wang, Q.; Yu, Y.; Yang, J.; Liu, J. Fast fabrication of flexible functional circuits based on liquid metal dual-trans printing. *Adv. Mater.* **2015**, *27*, 7109–7116. [CrossRef] [PubMed]

87. Votzke, C.; Daalkhaijav, U.; Mengüç, Y.; Johnston, M. Highly-stretchable biomechanical strain sensor using printed liquid metal paste. In Proceedings of the IEEE Biomedical Circuits and Systems Conference, Cleveland, OH, USA, 17–19 October 2018.

88. Gannarapu, A.; Gozen, B.A. Freeze-printing of liquid metal alloys for manufacturing of 3D, conductive, and flexible networks. *Adv. Mater. Technol.* **2016**, *1*. [CrossRef]

89. Liang, B.; Wei, J.; Fang, L.; Cao, Q.; Tu, T.; Ren, H. High-resolution rapid prototyping of liquid metal electronics by direct Writing on Highly Prestretched Substrates. *ACS Omega* **2019**, *4*, 21072–21077. [CrossRef] [PubMed]

90. Zhang, Q.; Gao, Y.; Liu, J. Atomized spraying of liquid metal droplets on desired substrate surfaces as a generalized way for ubiquitous printed electronics. *Appl. Phys. A* **2013**, *116*, 1091–1097.

91. Roberts, P.; Damian, D.D.; Shan, W.; Lu, T.; Majidi, C. Soft-matter capacitive sensor for measuring shear and pressure deformation. In Proceedings of the 2013 IEEE International Conference on Robotics and Automation (ICRA), Karlsruhe, Germany, 6–10 May 2013.

92. Jeong, S.H.; Hagman, A.; Hjort, K.; Jobs, M.; Sundqvist, J.; Wu, Z. Liquid alloy printing of microfluidic stretchable electronics. *Lab. Chip.* **2012**, *12*, 4657. [CrossRef]

93. Kim, K.; Choi, J.; Jeong, Y.; Cho, I.; Kim, M.; Kim, S.; Oh, Y.; Park, I. Highly Sensitive and Wearable Liquid Metal-Based Pressure Sensor for Health Monitoring Applications: Integration of a 3D-Printed Microbump Array with the Microchannel. *Adv. Healthc. Mater.* **2019**, *8*. [CrossRef]

94. Wong, R.D.P.; Posner, J.D.; Santos, V.J. Flexible microfluidic normal force sensor skin for tactile feedback. *Sensor. Actuat. A Phys.* **2012**, *179*, 62–69. [CrossRef]

95. Taekeon, J.; Yang, S. Highly stable liquid metal-based pressure sensor integrated with a microfluidic channel. *Sensors* **2015**, *15*, 11823–11835.

96. Zhang, L.J.; Gao, M.; Wang, R.H.; Deng, Z.; Gui, L. Stretchable pressure sensor with leakage-free liquid-metal electrodes. *Sensors* **2019**, *19*, 1316. [CrossRef]

97. Yang, Y.; Sun, N.; Wen, Z.; Cheng, P.; Lee, S.T. Liquid metal-based super-stretchable and structure-designable triboelectric nanogenerator for wearable electronics. *ACS Nano* **2018**, *12*, 2027–2034. [CrossRef] [PubMed]

98. Khoshmanesh, K.; Tang, S.; Zhu, J.; Schaefer, S.; Mitchell, A.; Kalantar-zadeh, K.; Dickey, M.D. Liquid Metal Enabled Microfluidics. *Lab Chip.* **2017**, *17*, 974–993. [CrossRef] [PubMed]

99. Gao, M.; Gui, L. Development of a fast thermal response microfluidic system using liquid metal. *J. Micromech. Microeng.* **2016**, *26*. [CrossRef]

100. Kramer, R.K.; Majidi, C.; Wood, R.J. Masked deposition of gallium-indium alloys for liquid-embedded elastomer conductors. *Adv. Funct. Mater.* **2013**, *23*, 5292–5296. [CrossRef]
101. Li, G.; Lee, D.W. Advanced selective liquid-metal plating technique for stretchable biosensor applications. *Lab. Chip.* **2017**, *17*. [CrossRef]
102. Lu, T.; Finkenauer, L.; Wissman, J.; Majidi, C. Rapid prototyping for soft-matter electronics. *Adv. Funct. Mater.* **2014**, *24*, 3351–3356. [CrossRef]
103. Pan, C.; Kumar, K.; Li, J.; Markvicka, E.; Herman, P.R.; Majidi, C. Visually imperceptible liquid-metal circuits for transparent, stretchable electronics with direct laser writing. *Adv Mater.* **2018**, *30*, 10–1002. [CrossRef]
104. Kim, T.; Kim, K.; Kim, S.; Lee, J.; Kim, W. Micropatterning of liquid metal by dewetting. *J. Microelectromech. Syst.* **2017**, *99*, 1–4. [CrossRef]
105. Ren, Y.; Wang, X.L.; Liu, J. Fabrication of High-Resolution Flexible Circuits and Sensors based on Liquid Metal Inks by Spraying and Wiping Processing. *IEEE Trans. Biomed. Circ. Syst.* **2019**, *1*, 1545–1551. [CrossRef]
106. Hong, S.Y.; Lee, Y.H.; Park, H.; Jin, S.W.; Jeong, Y.R.; Yun, J.; You, I.; Zi, G.; Ha, J.S. Stretchable active matrix temperature sensor array of polyaniline nanofibers for electronic skin. *Adv. Mater.* **2016**, *28*, 930–935. [CrossRef]
107. Jiao, Y.; Young, C.; Yang, S.; Oren, S.; Ceylan, H.; Kim, S. Wearable graphene sensors with microfluidic liquid metal wiring for structural health monitoring and human body motion sensing. *IEEE Sensor. J.* **2016**, *16*, 7870–7875. [CrossRef]
108. Hu, H.; Shaikh, K.; Liu, C. Super flexible sensor skin using liquid metal as interconnect. *IEEE Sensor.* **2007**, 815–817. [CrossRef]
109. Li, H.; Yang, Y.; Liu, J. Printable tiny thermocouple by liquid metal gallium and its matching metal. *Appl. Phys. Lett.* **2012**, *101*. [CrossRef]
110. Mohammed, M.G.; Kramer, R. All-printed flexible and stretchable electronics. *Adv. Mater.* **2017**, *29*. [CrossRef] [PubMed]
111. Otake, S.; Konishi, S. Integration of flexible strain sensor using liquid metal into soft micro-actuator. In Proceedings of the IEEE Micro Electro Mechanical Systems (MEMS), Belfast, UK, 21–25 January 2018.
112. Vogt, D.M.; Park, Y.L.; Wood, R.J. Design and characterization of a soft multi-axis force sensor using embedded microfluidic channels. *IEEE Sensor. J.* **2013**, *13*, 4056–4064. [CrossRef]
113. Ali, S.; Maddipatla, D.; Narakathu, B.B.; Chlaihawi, A.A.; Emamian, S.; Janabi, F.; Bazuin, B.J.; Atashbar, M.Z. Flexible capacitive pressure sensor based on PDMS substrate and Ga-In liquid metal. *IEEE Sensor J.* **2018**, *19*, 97–104. [CrossRef]
114. Cooper, C.B.; Arutselvan, K.; Liu, Y.; Armstrong, D.; Lin, Y.; Khan, M.R.; Genzer, J.; Dickey, M.D. Stretchable capacitive sensors of torsion, strain, and touch using double helix liquid metal fibers. *Adv. Funct. Mater.* **2017**, *27*. [CrossRef]
115. Zhang, R.; Ye, Z.; Gao, M.; Gao, C.; Zhang, X.; Li, L.; Gui, L. Liquid metal electrode-enabled flexible microdroplet sensor. *Lab. Chip.* **2019**, *20*, 496–504. [CrossRef]
116. Zhang, J.; Yao, Y.; Sheng, L.; Liu, J. Self-Fueled Biomimetic Liquid Metal Mollusk. *Adv. Mater.* **2015**, *27*, 2648–2655. [CrossRef]
117. Sivan, V.; Tang, S.; O'Mullane, A.; Petersen, P.; Eshtiaghi, N.; Kalantar Zadeh, K.; Mitchell, A. Enhanced electrochemical heavy metal ion sensor using liquid metal marbles - towards on-chip application. In Proceedings of the Optoelectronic & Microelectronic Materials & Devices, Melbourne, VIC, Australia, 12–14 December 2012.
118. Xu, J.; Fan, Y.; Yang, R.; Fu, Q.; Zhang, F. Realization of switchable EIT metamaterial by exploiting fluidity of liquid metal. *Opt. Express.* **2019**, *27*, 2837–2843. [CrossRef]
119. Ling, K.; Kim, K.; Lim, S. Flexible liquid metal-filled metamaterial absorber on polydimethylsiloxane (PDMS). *Opt. Express.* **2015**, *23*, 21375–21383. [CrossRef] [PubMed]
120. Kyeongseob, K.; Dongju, L.; Seunghyun, E.; Sungjoon, L. Stretchable metamaterial absorber using liquid metal-filled polydimethylsiloxane (PDMS). *Sensor* **2016**, *16*, 521.
121. Kenyu, L.; Hyung, K.; Minyeong, Y.; Sungjoon, L. Frequency-switchable metamaterial absorber injecting eutectic gallium-indium (EGaIn) liquid metal alloy. *Sensor* **2015**, *15*, 28154–28165.
122. Reichel, K.S.; Lozada-Smith, N.; Joshipura, I.D.; Ma, J.J.; Shrestha, R.; Mendis, R.; Dickey, M.D.; Mittleman, D.M. Electrically reconfigurable terahertz signal processing devices using liquid metal components. *Nat. Commun.* **2018**, *9*, 4203. [CrossRef]

123. Roychoudhury, S.; Rawat, V.; Jalal, A.H.; Kale, S.N.; Bhansali, S. Recent advances in metamaterial split-ring-resonator circuits as biosensors and therapeutic agents. *Biosens. Bioelectron.* **2016**, *86*, 595–608. [CrossRef]
124. Rodrigo, D.; Jofre, L.; Cetiner, B. Circular beam-steering reconfigurable antenna with liquid metal parasitics. *IEEE Trans. Antenn. Propag.* **2012**, *60*, 1796–1802. [CrossRef]
125. Mazlouman, S.; Jiang, X.; Menon, C.; Vaughan, R. A reconfigurable patch antenna using liquid metal embedded in a silicone substrate. *IEEE Trans. Antenn. Propag.* **2011**, *59*, 4406–4412. [CrossRef]
126. Hayes, G.; So, J.; Qusba, A.; Dickey, M.; Lazzi, G. Flexible liquid metal alloy (EGaIn) microstrip patch antenna. *IEEE Trans. Antenn. Propag.* **2012**, *60*, 2151–2156. [CrossRef]
127. Cheng, S.; Wu, Z.; Hallbjorner, P.; Hjort, K.; Rydberg, A. Foldable and stretchable liquid metal planar inverted cone antenna. *IEEE Trans. Antenn. Propag.* **2012**, *57*, 3765–3771. [CrossRef]
128. Oh, J.; Woo, J.; Jo, S.; Han, C. Pressure-conductive rubber sensor based on liquid-metal-PDMS composite. *Sensor. Actual. A Phys.* **2019**, *299*. [CrossRef]
129. Zhou, Y.; Wu, Y.; Asghar, W.; Ding, J.; Su, X.; Li, S. Asymmetric Structure Based Flexible Strain Sensor for Simultaneous Detection of Various Human Joint Motions. *ACS Appl. Electron. Mater.* **2019**, *1*, 1866–1872. [CrossRef]
130. Shi, X.; Cheng, C.H. Artificial hair cell sensors using liquid metal alloy as piezoresistors. In Proceedings of the IEEE International Conference on Nano/micro Engineered & Molecular Systems, Suzhou, China, 7–10 April 2013.
131. Liao, M.; Liao, H.; Ye, J.; Wan, P.; Zhang, L. Polyvinyl alcohol-stabilized liquid metal hydrogel for wearable transient epidermal sensors. *ACS Appl. Mater. Inter.* **2019**, *11*, 47358–47364. [CrossRef] [PubMed]
132. Kim, J.; Lee, M.; Shim, H.J.; Ghaffari, R.; Cho, H.R.; Son, D.; Jung, Y.H.; Soh, M.; Choi, C.; Jung, S.; et al. Stretchable silicon nanoribbon electronics for skin prosthesis. *Nat. Commun.* **2014**, *5*. [CrossRef] [PubMed]
133. Spina, F.; Pouryazdan, A.; Costa, J.; Ponce, C.L.; Münzenrieder, N. Directly 3D-printed monolithic soft robotic gripper with liquid metal microchannels for tactile sensing. *Flex. Print. Electron.* **2019**, *4*. [CrossRef]
134. Yi, L.; Li, J.; Guo, C.; Li, L.; Liu, J. Liquid metal ink enabled rapid prototyping of electrochemical sensor for wireless glucose detection on the platform of mobile phone. *J. Med. Devices.* **2015**, *9*. [CrossRef]
135. Feng, Y.; Huang, X.; Liu, S.; Guo, W.; Li, Y.; Wu, H. A self-powered smart safety belt enabled by triboelectric nanogenerators for driving status monitoring. *Nano Energy* **2019**, *62*, 197–204. [CrossRef]
136. Wang, H.; Xiang, Z.; Giorgia, P.; Mu, X.; Yang, Y.; Wang, Z.L.; Lee, C. Triboelectric liquid volume sensor for self-powered lab-on-chip applications. *Nano Energy* **2016**, *23*, 80–88. [CrossRef]
137. Wirekoh, J.; Valle, L.; Po, N.; Park, Y.L. Sensorized, flat, pneumatic artificial muscle embedded with biomimetic microfluidic sensors for proprioceptive feedback. *Soft Robot.* **2019**, *6*, 768–777. [CrossRef]
138. Lin, P.; Zhan, W.Z.; Yan, Q.; Xie, J.; Fan, Y.; Wu, M.; Chen, Y.; Cheng, Z. Capillary-based microfluidic fabrication of liquid metal microspheres toward functional microelectrodes and photothermal medium. *ACS Appl. Mater. Interfaces* **2019**, *11*, 25295–25305. [CrossRef]

Review

AC Electrothermal Effect in Microfluidics: A Review

Alinaghi Salari [1,2,3,*,†], **Maryam Navi** [1,2,3,†], **Thomas Lijnse** [4] **and Colin Dalton** [4,5,*]

1   Biomedical Engineering Graduate Program, Ryerson University, Toronto, ON M5B 2K3, Canada;
    maryam.navi@ryerson.ca
2   Institute for Biomedical Engineering, Science and Technology (iBEST), St. Michael's Hospital,
    Toronto, ON M5B 1T8, Canada
3   Keenan Research Centre, St. Michael's Hospital, Toronto, ON M5B 1T8, Canada
4   Biomedical Engineering Graduate Program, University of Calgary, Calgary, AB T2N 1N4, Canada;
    thomas.lijnse@ucalgary.ca
5   Electrical and Computer Engineering Department, University of Calgary, Calgary, AB T2N 1N4, Canada
*   Correspondence: a1salari@ryerson.ca (A.S.); cdalton@ucalgary.ca (C.D.);
    Tel.: +1-647-889-1276 (A.S.); +1-403-210-8464 (C.D.)
†   These authors contributed equally to this work.

Received: 11 October 2019; Accepted: 28 October 2019; Published: 11 November 2019

Abstract: The electrothermal effect has been investigated extensively in microfluidics since the 1990s and has been suggested as a promising technique for fluid manipulations in lab-on-a-chip devices. The purpose of this article is to provide a timely overview of the previous works conducted in the AC electrothermal field to provide a comprehensive reference for researchers new to this field. First, electrokinetic phenomena are briefly introduced to show where the electrothermal effect stands, comparatively, versus other mechanisms. Then, recent advances in the electrothermal field are reviewed from different aspects and categorized to provide a better insight into the current state of the literature. Results and achievements of different studies are compared, and recommendations are made to help researchers weigh their options and decide on proper configuration and parameters.

Keywords: electrothermal; microelectrode; microfluidics; micromixing; micropump

1. Introduction

Microfluidics is the precise control, and manipulation of fluids that are geometrically constrained to small, typically sub-millimeter, manufactured systems. Over recent decades, microfluidics has gained a great deal of attention in multiple fields, including medicine, chemistry, and biomedical engineering, due to its ability to perform multiplexing, automation, and high-throughput screening tasks. [1,2]. Due to the high surface-to-volume ratio of the fluid, and thus, the dominance of surface forces over inertial forces (i.e., low Reynolds number), fluid flow generation is a major challenge in microfluidic devices, as conventional pressure driven methods have poor efficiency in such devices [3,4]. The mechanisms of micro scale manipulation of fluids and particles can be categorized into two groups: mechanical, such as diaphragm-based devices, and non-mechanical, such as electrokinetic-based techniques. The presence of moving parts increases the risk of mechanical failure and can be incompatible with particulate flows, and thus, can limit the application of mechanical pumps for lab-on-a-chip devices [1,4,5]. Non-mechanical strategies, however, do not have these limitations. They can be integrated with microfluidic devices and also be used with particulate fluids [4,6,7]. Examples of non-mechanical methods include ultrasonic, direct current (DC) charge injection, and travelling wave driven electrohydrodynamic (EHD) micropumps [7–10].

Electrokinetics is a popular non-mechanical technique used for microfluidic fluid manipulation applications owing to its simple design and electronic automation [3]. Electrokinetic phenomena result from the interaction of an external electric field and induced electric charges. DC electrokinetics (DCEK), which has been studied over decades, requires relatively high voltages (i.e., on the order of several kilovolts) to operate which can limit its application in lab-on-a-chip devices [4,11,12]. alternating current (AC) electrokinetics (ACEK), however, which operates in low voltages (i.e., 1–20 V_{rms}), has led to the development of devices being portable and capable of handling biofluids without engaging in unwanted chemical reactions [13,14]. Furthermore, with non-uniform fluid flow streamlines generated by ACEK, this mechanism can be used to mix fluids [6].

AC electrokinetics mainly includes the dielectrophoresis (DEP), AC electroosmosis (ACEO), and AC electrothermal effects (ACET) [15], each of which is explained briefly in the following sections. There have been many substantive review articles on micropumps [16,17], electrohydrodynamics [18], electrokinetics [3,19,20], and their subcategories [2,21], but a comprehensive review focused on the electrothermal effect in microfluidics is still missing in the literature. This review intends to study the advances in the utilization of the electrothermal effect in microfluidics from different aspects, namely: the electric field, temperature field, and velocity field. In addition, channel properties, conditions of numerical simulations, experimental setup, and applications of ACET effect are presented. Finally, we will mention the ongoing research directions and future potential opportunities for the electrothermal effect. The majority of the publications cited throughout the manuscript that have made major contributions are also summarized in Table A1. It should be noted that the study of the electrothermal effect in other mechanisms such as ACEO is not in the scope of this paper and can be found elsewhere [5,22]. Furthermore, strategies which are based on DC electric fields (e.g., DC electrophoresis) or non-electrical (e.g., magnetic) forces are not discussed here.

2. AC Electrokinetics

2.1. Dielectrophoresis

Dielectrophoresis arises from the interaction between a dipole moment on a particle and a non-uniform electric field [23]. If the particle has a polarizability higher than the surrounding medium, the DEP force exerting on the particle will be towards regions with a high electric field (positive DEP). For particles with lower polarizability, however, this force will be towards regions with a low electric field (negative DEP). This is demonstrated by the Clausius–Mossotti factor, which specifies the direction of the DEP force with respect to the electric field [24]. In addition to the permittivities of the particle and the medium, the magnitude of DEP force is also a function of the particle size. DEP force is directly related to the third power of the particle radius, and thus, is an ideal tool for separating, concentrating, and sorting particles, cells, and viruses [25–31]. Moreover, since DEP force scales with the gradient of the electric field, it decreases with the distance from electrodes, and the generated velocity is inversely proportional to the third power of distance [13]. Therefore, DEP is not an effective technique for handling particles of relatively small size (e.g., ≤ 1 μm) far from a strong electric field (e.g., a few micrometers away) [32]. Two excellent reviews on dielectrophoresis are [33,34].

2.2. AC Electroosmosis

AC electroosmosis is dependent on the formation of an electric double layer (EDL) at the interface of a liquid and solid substrate [35–37]. At an interface of a solid object and electrolyte fluid, due to the adsorption of ions onto the object surface it acquires charges, and as a result, an EDL forms inside the fluid near the surface [3,38]. When an electric field is applied to this system, the charges in this layer experience an electrostatic force, which can cause fluid motion. The rest of the fluid is then dragged into motion due to viscous forces.

Since the EDL thickness is inversely related to the fluid electrical conductivity, at relatively high conductivities (e.g., 84 mS·m^{-1}), the EDL thickness becomes very small (e.g., <1 nm), which makes ACEO ineffective for the manipulations of biological fluids (1–2 S·m^{-1}) [1,13,15,39–41]. In addition, ACEO is frequency-dependent, and increasing the actuation frequency beyond 100 kHz causes the

ACEO effect to become invisible, since at high frequencies, the electric double layer is unable to form, and no fluid flow is generated [35]. Similarly, at very low frequencies, the double layer can completely screen the electric field, and thus, no net flow can be generated. ACEO has been developed and used in many forms to pump fluids or manipulate particles, namely biased ACEO for particle assembly and micropumping [32,42], micropumping of fluids [43,44], Travelling wave ACEO [45] and asymmetric ACEO micropumping [40], and DEP electrohydrodynamic particle trapping (i.e., ACEO in conjunction with DEP) [46,47]. As ACEO is only effective in relatively low frequencies, it is more prone to bubble generation and electrode deterioration resulting from electrochemical reactions, which can affect the electric field distribution and eventually damage the device [39]. Despite these limitations, there are many application-driven papers in the literature using ACEO [48–51], where modifications have been suggested to enhance ACEO applicability in fluids with electric conductivities up to 0.1 S·m^{-1} under the actuation of high frequencies and voltages. These modifications include the utilization of polarizable walls in induced-charge electroosmosis [52], AC faradic polarization [50], and nonlinear electroosmosis on curved surfaces [53], to name a few. An advantage of ACET is that it can be used for higher conductivities, i.e., over 1 S·m^{-1}. More details of different strategies for implementation of ACEO in microfluidics can be found in [54,55]. DEP and ACET effects can be combined [13,56–59] to improve particle manipulations, as DEP has difficulty in manipulating submicron particles, and also is weak in areas far from electrodes where ACET is strong [13,56].

2.3. AC Electrothermal

Unlike ACEO and DEP, ACET has been shown to be very effective in biomedical applications which involve high conductivity biofluids, such as blood, urine, and saliva [60]. This is due to the fact that the ACET effect originates from a temperature gradient in the bulk of the fluid and not the fluid-electrode interface (i.e., the EDL). Fluids with higher conductivities can generate stronger microflows, and therefore, it is the most efficient electrokinetic mechanism for manipulating biological fluids with conductivities above 0.7 S·m^{-1} [1,4,13,14,39,60–63].

Emerging in 1960s [64,65], ACET has been widely used for fluid manipulations over the years [7,66], and is also referred to as induction EHD [67,68]. Despite similar flow patterns, the physics behind ACET and ACEO are different. ACEO is the result of the interaction of an electric double layer at the interface of the fluid-electrode and a non-uniform AC electric field, while ACET arises from the interaction of a temperature gradient in the bulk of the fluid and a non-uniform AC electric field. The source of the temperature gradient may be internal (i.e., Joule heating) or external (e.g., strong illumination, microheaters, etc.). Temperature gradients in the fluid lead to gradients in the electrical properties of the fluid, i.e., conductivity and permittivity, which induce free charge density. An electric force arising from the non-uniform electric field causes the free charges to move. As a result of shear stress, the surrounding fluid is also dragged into motion which produces microflows. Unlike ACEO, the ACET effect shows plateaus in force in a wide frequency range (10–10^{11} Hz) [39,69]. With ACET, the fluid velocity is steadier and more predictable at different frequencies compared to ACEO and DEP. In general, ACET flow can be generated in frequencies above 100 kHz and salt concentrations of above 10^{-2} mol·dm^{-3}, whereas ACEO is more common at low frequencies and salt concentrations of 10^{-2} mol·dm^{-3} and below [62]. Despite the fact that fluid heating is crucial for the ACET effect, ambient heat conduction helps dissipating energy so that the temperature rise in the bulk of the fluid is typically low ($\Delta T < 5$ K), which is safe for biofluids [11,13,14,41]. The ACET force is proportional to the temperature gradient $|\nabla T|$ and not the temperature rise [1].

In order to generate AC electric fields required for inducing the electrothermal effect, microfabricated electrode arrays are commonly used. Employing a symmetric pair of electrodes at the bottom of a microfluidic channel can induce two symmetric sets of microvortices above the electrodes, and thus, no net flow can be generated [70]. For pumping applications, however, the electrode symmetry needs to be broken. Since the electrothermal force is a function of the electric field and temperature gradient, asymmetry may be achieved by manipulating either or both of these factors. This will be discussed in more details in the following sections. Typically, due to its simple implementation, imposing geometry asymmetry to microelectrodes is the most common approach for breaking the symmetry of microvortices. In addition, manipulating the temperature field with the help of external heat sources, such as strong illumination [69,71–73], embedded microheaters [74,75], and heating electrodes [1], can also be used for creating a net flow. Although a common ACET microdevice implements an array of electrode pairs placed at the bottom of a microchannel with a rectangular cross section, more complicated configurations with electrode arrays placed on the top, bottom, and sidewalls of channels with different cross sections have also been studied [69,76–78]. Studies with the use of grooves on the channel surface to induce further asymmetry and increase flow have also been addressed, but fabrication of these designs suffers from serious challenges.

Similar to other electrokinetic mechanisms, ACET suffers from some drawbacks, most of which have been addressed to some extent in the literature, as will be shown in this paper. In microfluidic devices, miniaturization can be hindered as the ACET effect originates from the bulk of the fluid and decreasing the channel dimensions can decrease the volume of the fluid flowing inside the channels [3,35]. In addition, ACET depends on the formation of temperature gradients, and therefore, cannot be used with low conductivity fluids. As such, its application in conjunction with DEP, which requires low conductivity fluids for efficient particle sorting, is limited [1,4,5]. Importantly, an excessive temperature rise in fluids with high conductivities can cause the buoyancy force to dominate over the ACET force [4]. The reason is that the ratio of electrothermal force to buoyancy force is proportional to $|\nabla T|/\Delta T$. Thus, when $\Delta T > |\nabla T|$, the buoyancy force becomes the dominant force. Finally, increasing the actuation voltage to achieve high fluid velocity can lead to electrochemical reactions which can limit the application of ACET effect on biofluids [1]. There have been some reports on how to mitigate this issue [79].

3. Theory

As stated in the previous section, the ACET effect results from the interaction of a non-uniform electric field and a temperature gradient in the bulk of the fluid. The energy balance equation governs the amount of Joule heating as follows [15]:

$$k\nabla^2 T + \frac{1}{2} < \sigma |E|^2 >= 0 \tag{1}$$

where, k and σ are the thermal and electrical conductivities of the fluid, respectively, and E is the electric field, which can be obtained from the Laplace equation in a homogeneous medium as below:

$$\nabla^2 V = 0 \tag{2}$$

where, $E = -\nabla V$, and V represents the electric voltage.

An order of magnitude estimation of Equation (1) gives [15]:

$$\Delta T \approx \frac{\sigma V^2}{k} \tag{3}$$

Based on Equation (3), the temperature rise is directly proportional to the fluid electrical conductivity and actuation voltage squared, which, in most applications, is the control parameter.

The ratio of heat convection to heat conduction in a microchannel is very low (i.e., Peclet ≤ 0.07) [15,67,80]. Furthermore, it has been numerically shown that, compared to electrical forces, natural convection in micro-channels can be neglected [15,80]. However, for cases with high thermal Peclet numbers, heat convection cannot be neglected [81]. The temperature gradient in the fluid can change the fluid properties, including permittivity ε and conductivity σ, and can be calculated as follows [15]:

$$\nabla\varepsilon = \left(\frac{\partial\varepsilon}{\partial T}\right)\nabla T \tag{4}$$

$$\nabla\sigma = \left(\frac{\partial\sigma}{\partial T}\right)\nabla T \tag{5}$$

In most ACET applications, it is assumed that the rate of change of permittivity and conductivity with the change in temperature is very small [15]. Otherwise, a temperature coefficient needs to be defined to account for the changes in fluid properties [81,82]. As a result of this assumption, the perturbed electric field can be neglected, and the charge convection can be assumed to be much smaller than the charge conduction [15].

The change in electrical properties of the fluid leads to the generation of electrical charge density as follows [15]:

$$\rho_E = \nabla\cdot(\varepsilon E) \tag{6}$$

$$\frac{\partial\rho_E}{\partial t} + \nabla\cdot(\sigma E) = 0 \tag{7}$$

where, ρ_E is the charge density.

Under the effect of the electric field, there is a force applied to the charge density which is [15]:

$$f_E = \rho_E E - \frac{1}{2}E^2\nabla\varepsilon \tag{8}$$

The first term in Equation (8) is the Coulomb force and the second term is the dielectric force.

As charges move in the electric field, they drag the surrounding medium into motion. Therefore, microflows are generated in the fluid and are governed by:

$$\nabla p + \eta|\nabla|^2 u + f_E = 0 \tag{9}$$

where, η, p, and u are the dynamic viscosity, pressure, and velocity field, respectively. Furthermore, from the conservation of mass for an incompressible fluid, we have:

$$\nabla\cdot u = 0 \tag{10}$$

With an order of magnitude estimation from Equation (9), the flow velocity can be written as $|u| \approx< f_E > \cdot\frac{l^2}{\eta}$, where l is the characteristic length of device, which is usually the electrode spacing [4,15,60].

Charge density can be calculated by combining Equations (6) and (7) as follows [83]:

$$\rho_E = \frac{\sigma\varepsilon}{\sigma + i\omega\varepsilon}(\alpha - \beta)(\nabla T\cdot E) \tag{11}$$

where, ω is the angular frequency of the AC electric field, and:

$$\alpha = \frac{1}{\varepsilon}\left(\frac{\partial\varepsilon}{\partial T}\right) \tag{12}$$

$$\beta = \frac{1}{\sigma}\left(\frac{\partial\sigma}{\partial T}\right) \tag{13}$$

For aqueous solutions and temperatures around 293 K, α and β can be estimated as -0.4% K^{-1} and 2% K^{-1}, respectively [84].

With the above approximations, the electrothermal force can be simplified as below [15]:

$$< F_E > = \frac{1}{2}\frac{\varepsilon(\alpha-\beta)}{1+(\omega\tau)^2}(\nabla T \cdot E)E - \frac{1}{4}\varepsilon\alpha|E|^2\nabla T \tag{14}$$

where, $\tau = \frac{\varepsilon}{\sigma}$ is the charge relaxation time of the liquid and is in the range of 0.7–35 ns for conductivities in the range of 0.02–1 S·m^{-1} [41,85]. As stated above, the first term represents the Coulomb force, and the second term is the dielectric force. These forces act in different frequency ranges (i.e., the Coulomb force dominates at low frequencies and dielectric force dominates at high frequencies) and are in different directions [83]. Near a certain frequency, known as the cross-over frequency f_c, the two forces compete, and flow reversal can occur as a result of switching from a Coulomb force dominant to a dielectric force dominant regime or vice versa [15]. The cross-over frequency can be calculated as below [15]:

$$f_c \approx \frac{1}{2\pi\tau}\left(2\frac{\left|\frac{1}{\sigma}\left(\frac{\partial\sigma}{\partial T}\right)\right|}{\left|\frac{1}{\varepsilon}\left(\frac{\partial\varepsilon}{\partial T}\right)\right|}\right)^{\frac{1}{2}} \tag{15}$$

For example, the cross over frequency for a biofluid with a conductivity of 1 S·m^{-1} is roughly 200 MHz [61]. ACET force, and thus ACET flow velocity, is higher (up to 11 times) at low frequencies and has no dependence on frequency except for the transition region (i.e., near crossover frequency) [13,41,69].

By taking a closer look at Equation (11), as α, β, and ω are constants, we can conclude that $\rho_E \propto \nabla T \cdot E$ [1]. Commonly, in electrokinetics, frequencies much lower than 10 MHz (usually around 200 kHz) are used, where $1 + (\omega\tau)^2 \approx 1$ and dielectric force is negligible (i.e., the Coulomb force is ~11 times larger than the dielectric force) [41]. At these frequencies, there is not enough time for the double layer to form, and thus, the dielectric force is neglected [13]. As a result, the flow direction is determined by the Coulomb force and Equation (14) is reduced to the first term on the right hand side. With a similar argument, as α, β, and $\omega\tau$ are constants, we can conclude that $|< F_E >| \propto |E|^2|\nabla T|$, and according to Equation (1), when Joule heating is implemented, $\nabla T \propto E^2$, and therefore, $|< F_E >| \propto |E|^4$ [69]. This means that electrothermal force is proportional to the fourth power of the electric field. According to the order of magnitude estimation of velocity obtained from Equation (9), ACET fluid velocity also has a quartic relationship with the actuation voltage. This means that a small increase in the electric field can cause a large increase in electrothermal force, and thus, a significant increase in the resultant flow velocity. As stated above, since electrothermal and electroosmotic flows have similar patterns, one way to distinguish them is to use this proportionality. Since electroosmotic velocity is proportional to the square of voltage, by plotting velocity against applied voltage, the source of microflows can be revealed [14,15].

The theory discussed here is based on an uncoupled model developed by Ramos et al. in 1999 [15]. In such a model, an assumption of a small temperature rise $\Delta T < 5$ K renders the change of fluid properties and electric field with temperature insignificant. However, if the temperature rise is considerably higher, then the fluid properties will change by temperature variations, and therefore, a fully coupled model needs to be used [74,86–88]. Hong et al. developed a coupled model, compared it with the classical model, and found that only at small temperature rises do the results of the two models match [87]. More recently, pumping and mixing of non-Newtonian fluids have been studied [89–91].

4. Electric Field

In lab-on-a-chip applications, the maximum limit for actuation voltage is typically within the range of 5–7 V$_{\mathrm{rms}}$, beyond which electrochemical reactions can harm the nature of biofluids. Since, as stated in the previous section, electrothermal force, and thus, fluid velocity is highly dependent upon the applied voltage (i.e., quartic dependence), the increase in velocity of electrothermal based biomedical microdevices is hindered accordingly. To overcome this issue, many studies have proposed and developed new ACET designs with modifications to the electric field, which include either changes to the geometry of the electrode array or introducing asymmetry to the electric potential, using techniques such as travelling wave, multiphase, DC biased, etc. In this section, different strategies proposed for modifying the electric field are reviewed.

4.1. Introducing Asymmetry to Geometry

To produce a strong electric field, microelectrodes are typically patterned on the inner surfaces of microchannels [11,66]. Due to changes to the electric field and temperature gradient strengths, by manipulating the electrode configuration, a wide variety of flow patterns and velocities can be achieved [41]. Microelectrode arrays with different shapes have been reported, which can be categorized into two-dimensional (2D) and three-dimensional (3D) geometries [41,85].

Due to the simplicity and ease of fabrication, 2D electrode geometries are the most common and can be further categorized into asymmetric rectangular [4,41,69,92], orthogonal [4], meandering [62], concentric [60], and triangular [56], with asymmetric rectangular being investigated the most in the literature (Figure 1a–e). Imposing asymmetry on the electrode pairs induces asymmetry in the resultant microflows, and therefore renders pumping action. Due to this feature of microelectrode configuration, more heat is hypothesized to be generated near the narrow electrode [1]. In general, in such a configuration, the width of the narrow electrode and the gap between electrodes in a pair are the two major design parameters which govern the magnitude of the fluid velocity [41].

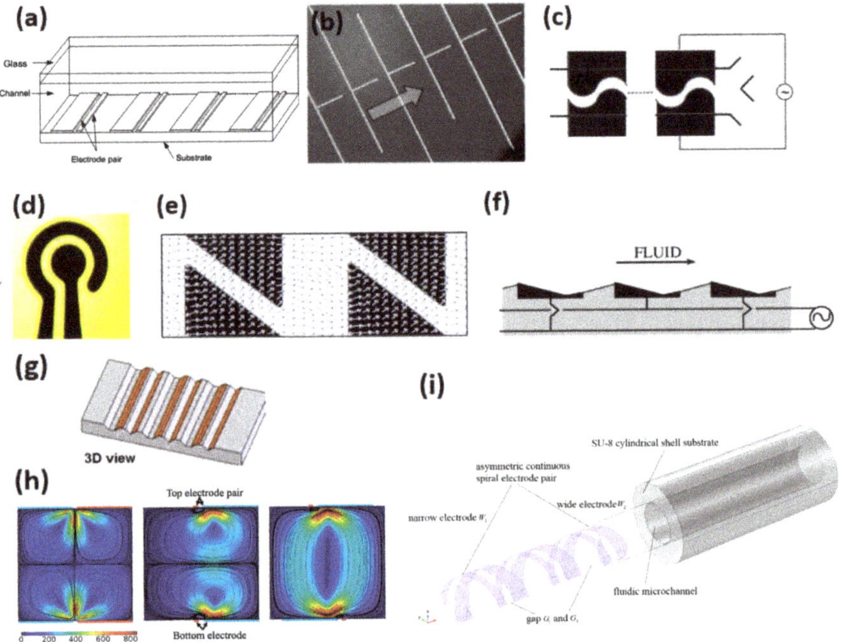

Figure 1. Schematic of different electrode geometries. (**a–e**) 2D Electrodes: (**a**) asymmetric rectangular, (**b**) orthogonal, (**c**) meandering, (**d**) concentric, and (**e**) triangular. (**f–i**) 3D Electrodes: (**f**) 3D electrode array, (**g**) microgrooved configuration, (**h**) electrodes facing each other, and (**i**) spiral design. Reproduced with permission from [56,60,62,69,77,85,93–95].

Yuan et al. conducted a thorough study on the optimization of 2D rectangular electrode arrays and reported a set of ratios for the corresponding parameters [41]. 2D rectangular asymmetric arrays are used mostly for pumping applications, either alone or with other changes to electric or thermal fields [1,74]. In pumping applications, one or multiple arrays of asymmetric microelectrode pairs are placed perpendicular to the channel length in order to generate a net flow along the channel. In applications, where a lateral fluid mixing is also desirable, however, the electrode array needs to be placed at an angle <90° to the channel length [77,95]. In 2D electrode configurations, by decreasing the gap between electrodes up to a certain point, the resultant fluid velocity can be increased as the strength of the electric field increases. However, if the strength of the electric field is maintained at a constant, increasing the gap between electrodes can enhance the resultant fluid flow, as increasing the gap allows a larger volume of fluid bulk above the gap to experience the strong electric field [41]. One way to enhance the ACET effect is to decrease the width of the narrow electrode, increasing the non-uniformity of the electric field. However, if the narrow electrode is made too small, the strength of the electric field can decrease accordingly, and, as electrothermal force is proportional to the fourth power of the electric field, electrothermal force and, hence, velocity can decrease drastically [41]. As shown, the dimensions and electrode geometry are of great importance to the electrothermal effect. In addition, the channel size and cross section can also significantly affect the resultant flow rate [96]. More on this topic can be found in Section 9. Orthogonal electrode arrays were proposed by Wu et al. for pumping applications with velocities exceeding 1 mm·s^{-1} [4,93]. A meandering electrode array, with a sinusoidal electrode gap, was proposed for rapid mixing of two biofluids [62]. Rapid mixing of high-conductivity (up to 22 S·m^{-1}) fluids has been shown using concentric electrode designs [60]. A triangular electrode pattern has also been suggested for micromixing [56].

The backward flow generated above the narrow electrode can negatively impact the use of planar configurations of electrodes for pumping applications, as they cause the average flow in the microchannel to slow down [85]. In 2000, Ajdari, suggested a 3D electrode configuration to circumvent this problem [94] (Figure 1f). Expanding on the same idea, Du and Manoochehri suggested a microgrooved channel (instead of 3D electrodes) to impose spatial asymmetry to the electrode pairs [97] (Figure 1g). More recently, this effect has been numerically simulated with a number of substrate configurations [98]. Manufacturing a channel with modified roughness of substrate surface is easier than manufacturing 3D electrodes. In their work, different shapes of grooves are proposed and experimentally studied. This study showed that by using the microgrooved structure, based on the shape of the grooves, the pumping capacity increases by up to six-fold compared to a planar configuration. In another work, they reported an optimized configuration of such a design [85]. They managed to increase the net flow rate by further suppressing backflows and shortening the streamlines [85]. However, fabrication of such microgrooved structures is more complicated than conventional microchannels with planar electrode design [69]. A variation of such configuration in conjunction with opposing electrodes on the top surface is suggested for mixing applications [89,99]. A similar configuration is also reported where opposing castellated electrodes are utilized to eliminate the ACET vortex and thereby enhance pumping [100].

A two-layer microelectrode configuration, where microelectrodes are facing each other, has been proposed for particle trapping in order to investigate the DEP effect in conjunction with ACET [13]. In this electrode configuration, one electrode is placed on top and the other at the bottom of the channel. A similar configuration is also used for studying Joule heating effects [80], as well as for patterning of colloids when equipped with underlying microheaters [75]. Such an opposing microelectrode configuration with different patterns is also reported for mixing fluids [77,89,101] (Figure 1h). This configuration is also used in rapid electrokinetic patterning, where electrothermal effects play a significant role in particle manipulation [102,103]

More recently, studies have been carried out on utilizing electrodes on all walls of the microchannel in different patterns [104,105]. Using a particular configuration of multiple electrode arrays on side walls, simultaneous mixing and pumping of biofluids in one microchannel during a short time and over a short distance is feasible [104]. In a very recent publication, a spiral electrode pattern has been proposed for simultaneous mixing and pumping which is capable of achieving a velocity of 400 μm·s^{-1} for a biofluid with a conductivity of 0.224 S·m^{-1} (Figure 1i) [95].

4.2. Introducing Asymmetry in Electric Potential

In addition to an asymmetric electrode geometry, the electric field can also be modified to enhance the electrothermal effect. Applying a DC potential to the electrode pairs in addition to the existing AC potential, using a multiphase AC signal in the form of a travelling wave, and utilizing a two-phase actuation system are examples of such modifications, each of which is briefly described here.

4.2.1. DC Biased

In this method, symmetric electrodes are used but one electrode is always at a positive potential and the other is always at a negative potential subject to faradaic charging and capacitive charging, respectively (Figure 2a) [6]. Hence, positive charges on both electrodes lead to unidirectional flow. The advantage of this configuration is that pumping action can be achieved by a symmetric pair of electrodes.

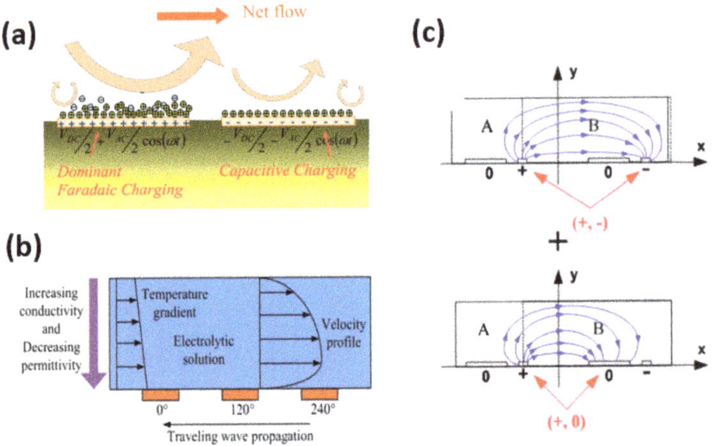

Figure 2. Schematic of different mechanisms for imposing asymmetry to the electric field in alternating current electrothermal effects (ACET) devices. (**a**) Direct current (DC) biased configuration. The left electrode is always at a positive potential and subject to faradaic charging, while the right electrode is always at a negative potential and subject to capacitive charging. (**b**) Travelling wave configuration. The induced temperature gradient generates a charge density which is moved with the travelling electric field. Note that the direction of propagation of travelling wave is opposite to the direction of fluid flow. (**c**) Two-phase asymmetric configuration. The narrow electrodes in two adjacent pairs have different polarities. The superposition of the two configurations enhances the electric field in region B. Reproduced with permission from [6,68,69].

4.2.2. Travelling Wave (TW)

For the first time in 1966, Melcher observed pumping of fluids with very small conductivities, by imposing a temperature gradient along the depth of the channel in conjunction with travelling wave induction [64]. In 1967, Melcher and Firebaugh further investigated the phenomenon and presented the governing equations [65].

Fuhr et al. studied a travelling wave induced micropump and modified the principles presented by Melcher and Firebaugh [7,65]. Later, in 1994, the same group explained the mechanisms in which temperature gradients (both externally applied and internally generated) are used in conjunction with a travelling wave [66]. With an induced temperature gradient in the fluid bulk, free charges can be generated, and the travelling wave can move them along the channel above the microelectrodes causing fluid flow. This principle is further illustrated in Figure 2b. The mechanism of applying a temperature gradient in this design can be external (i.e., bed heating) or self-generated (i.e., Joule heating).

In 1992, Fuhr et al. experimentally investigated a high frequency travelling wave induced micropump with self-generated temperature gradient, i.e., higher temperatures at the bottom of the channel in the space between electrodes and lower temperatures at the top of the channel [7]. In this case, the resultant fluid flow is in the opposite direction of the travelling wave, as also verified by Liu et al., [68]. Their measurements of flow velocity show a quadratic relationship with voltage caused by applying an external temperature gradient [13,106]. Following the study of Fuhr et al., several numerical investigations of travelling wave induced electrothermal flow were also carried out [68,107].

4.2.3. Two-Phase Actuation

In 2011, Zhang et al. proposed a two-phase planar asymmetric electrode design which yielded an increase of 25–50% in flow rate compared to conventional single-phase asymmetric configurations [69]. In this design, an AC signal of 0°/180° was applied to the narrow electrode in an asymmetric electrode pair (Figure 2c). This led to an enhanced electric field. As the temperature gradient has a quadratic relationship with the electric field, temperature, and therefore, flow rate increased accordingly.

5. Temperature Field

Internal and External Heating

In an ACET device, heating of the fluid is the source of thermal gradients which can eventually lead to electrothermal microflows. Heating can be either internal (i.e., Joule heating) or external.

Internal heating refers to Joule heating of the fluid upon activation of the electric field. In a rectangular asymmetric electrode configuration, the temperature rise above the narrow electrode is higher than that above the wide electrode. Joule heating scales with conductivity of the fluid and applied voltage (Equation (1)). As a result, fluids with low conductivity need a higher electric field to produce sufficient thermal gradients. However, at high applied voltages, electrochemical reactions can occur, which can damage the fluid and electrodes. This issue can be resolved by applying an external heat source and keeping the applied voltage low.

A few methods have been proposed for external heating, including strong illumination [69,71,73], integrated heating elements (ohmic heating element) [39], heating of the electrodes [1], and thin film resistive heaters [74], the latter of which is shown to be portable and more efficient [74,75].

Incorporation of external heating allows the electrothermal flow to be controlled independently of the electric field strength [39,74]. Therefore, the effect of fluid properties on the resultant flow is minimized [39], and manipulation of low conductivity fluids with the electrothermal effect becomes feasible [74]. In addition, by means of external heating, the electrothermal flow can be implemented in conjunction with other electrokinetic methodologies, which work under a uniform electric field, such as electrophoresis and electroosmosis [74,75]. External heating strategies also enable control of the direction of fluid flow [1,71,106].

Green et al. performed a thorough study on strong illumination as an external heat source [71,106]. The resultant velocity field in the case of external heating can be obtained from the equation below [13,106]:

$$|u| \approx 3 \times 10^{-3} \left(\frac{\varepsilon V^2}{\eta} \right) \left| \frac{\partial T}{\partial y} \right| \left(\frac{1}{\sigma} \right)$$

(16)

where, $\frac{\partial T}{\partial y}$ is the external thermal gradient along axis y. Unlike Joule heating, here the velocity has a quadratic relationship with voltage, which was verified experimentally by Stubbe et al., where ohmic heating elements were used [39].

Depending on the amount of heat flux introduced into the fluid bulk by the external heating, the effect of voltage on the flow velocity changes. In 2013, Yuan et al. found that if the heat flux is

Micromachines **2019**, *10*, 762

small (around 10^4 W·m^{-2}), it can be counteracted by the Joule heating of the fluid, and the relationship between fluid velocity and voltage can be between quartic and quadratic [1]. However, if the heat flux is relatively large and the fluid is mostly under the influence of external heating, the fluid velocity will have a quadratic dependence on the voltage.

Furthermore, thin film microheaters can be embedded in the substrate below the electrodes, separated by an insulating layer [75]. Such a heating mechanism can yield a thermal efficiency of close to 100% [74]. Velasco and Williams showed the application of thin film heaters for assembly of colloids [75]. The geometry of the assembly is governed by the geometry of the array of microheaters. Therefore, particles can be trapped in a larger area when compared to conventional particle trapping techniques [13]. The assembly of 1 µm and 2 µm particles was achieved in the frequency range of 1–200 kHz, while no particle aggregation was observed at other frequencies. Since the temperature gradient is the underlying mechanism, microheaters cannot be patterned close to each other. Furthermore, with thin film microheaters, sharper temperature gradients exist near the electrodes which result in a larger value of $\mathbf{V}T{\cdot}E^2$ [74]. Compared to Joule heating, thin film microheaters alone require only 40% of the power to produce the same flow rate. Without a change in power, the flow rate can be increased to 250% of that of solely Joule heating [74]. Both the proximity of thin film resistive heaters to each other and their distance to the regions of maximum electric field strength determine the resultant flow regime [74,75]. The maxima of temperature gradient and electric field need to occur in the same region in the bulk of the fluid in order to maximize the product of $\mathbf{V}T{\cdot}E^2$ [74]. More recently, Williams and Green applied the same idea to a symmetric pair of electrodes, which is more desirable for DEP applications, and carried out numerical simulations towards finding an optimum location for the heater with respect to the electrodes [108].

In spite of a fundamental factor for generation of electrothermal flow, Joule heating can be an unwanted effect in other electrokinetic devices [80]. For example, in a DEP based particle manipulation device, Joule heating can harm the biological fluid and form electrothermal microflows taking particles away from their intended spot of sorting or trapping [80,81]. For this reason, studies have been conducted to gain a better insight into Joule heating [80] and its effects on insulator-based DEP devices [81,109].

Almost all electrothermal designs studied in the literature are checked for excess temperature rise due to Joule heating, as this can negatively impact their potential application with biofluids. For example, Du and Manoochehri compared the Joule heating in their proposed microgrooved and planar configurations at a wide range of conductivities [85] and found no major difference between the two structures. In devices with thin film microheaters, temperature rise is shown to be half of that of Joule heating, which further corroborates the high efficiency of these devices [74].

Similar to Joule heating, external heating may not always lead to an enhanced electrothermal effect. Zhang et al. studied the effect of strong illumination coupled with their two-phase planar electrode configuration [69]. Interestingly, they observed a decrease in velocity when a strong illumination was applied. This observation was justified by assuming that the flow direction generated by the illumination is opposite to that generated by Joule heating [83]. As mentioned earlier, for external heating to be effective, heat flux introduced into the system must be significantly higher than the Joule heating generated by the system. Not clearly stated by Williams [74], however, is that their reported results show that by increasing the conductivity, the ratio of electrothermal force generated by thin film heaters to that generated by Joule heating gradually decreases. This indicates that the coupled electric and temperature fields need to be solved for higher temperature rises (>5 K) [82,86].

ACET flow velocity has a strong dependence on the applied voltage ($u \propto V^4$, when no external heating is applied). However, voltage can be increased up to a certain point, usually below 7 V$_{\text{rms}}$, to avoid thermal damage to the biofluid [1,11,14,41,60,71,75]. To solve this problem, Yuan et al. introduced a thermally biased configuration, where one electrode in a pair is at a higher temperature than the other one [1]. In this way, the temperature gradient in the fluid can be controlled independently from the voltage.

At frequencies below 10 MHz, where the Coulomb force is the dominant force, the electrothermal force can be simplified to $F_E = \rho_E E$. Since, due to the limitations discussed above, there is an upper limit for the applied electric field, the electrothermal force can be increased further by increasing the charge density, ρ_E. As $\rho_E \propto \nabla T \cdot E$, Yuan et al. attempted to increase the electrothermal force by increasing the temperature gradient by means other than Joule heating [1]. In their design, unidirectional flow was obtained with symmetric electrodes, making it possible for their configuration to be used either as a micromixer or a micropump depending on the applied voltage, frequency, and heat flux. When imposing external heat flux to the narrow electrode in an asymmetric pair, they obtained a velocity 5.7 times higher than that of a regular asymmetric ACET micropump. Their simulation results show that by applying the heat flux to the wide electrode, the direction of the flow above the electrodes can be reversed. This finding corroborates their hypothesis that the generated heat on the narrow electrode is the primary cause of pumping in a conventional ACET device, and that external heating has a strong influence on the pumping performance.

6. Fluid Flow Regime

6.1. Flow Velocity

Almost all the work that has been performed dealing with the electrothermal effect is an attempt to increase the ACET velocity in high conductivity fluids while avoiding significant increase in temperature and voltage. Many studies have attempted to increase the electrothermal flow velocity by focusing on the governing parameters mentioned in Equation (14) and have proposed new designs to increase the velocity based on their interpretations of the formula. The maximum of the flow velocity reported in these studies differs both in its magnitude and location (i.e., height/distance from channel bottom/electrodes).

Since electrothermal velocity has a quadratic relationship with the temperature gradient, many studies have focused on increasing this parameter of electrothermal force to increase the flow velocity. External heating has been proposed and thoroughly investigated for this purpose, which is discussed in Section 5. By simplifying the equation of electrothermal force to $|< F_E >| = \xi(\omega)|E|^2|\nabla T|$, Zhang et al. concluded that by increasing both the strength of electric field and the temperature gradient, a significant increase in electrothermal force and thus fluid velocity could be obtained [69]. Taking advantage of this combination, by developing a two-phase asymmetric planar electrode design in which a stronger electric field is obtained, they demonstrated flow velocities reaching 25–50% higher than those of conventional single-phase configurations. In this simplified equation, $\xi(\omega)$ is a function of frequency and the angle between the vectors of applied field and temperature gradient. They also found that fluid velocity is much higher at low frequencies, which was also verified by Williams who determined that pumping rates at high frequencies ($>f_c$) drop to 10% of that at low frequencies [74].

To accurately measure the electrothermal velocity field, usually micro particle image velocimetry (micro-PIV) is used. The assumption of this method is that particles follow the fluid flow and are not under the influence of any other forces [110–112]. Therefore, ACET velocity needs to be measured at a height with a negligible DEP effect on particles [1]. At a distance close to electrode surface, the DEP effect on particles is significantly high, which can cause them to be trapped at the electrode edges [1]. As a result, depending on the channel height and size of tracer particles, ACET velocity is typically measured at a height of 10–50 μm above the electrode surface. A good control for this effect is carried out by Wu et al., where, they found, by using particles with a size of ~500 nm, an order of magnitude estimation yields that at ~5 V_{rms} particles move with a DEP-induced flow velocity of 0.11 μm·s^{-1} at the height of 10 μm above the electrodes [4]. Since ACET velocity is often higher than 100 μm·s^{-1}, DEP velocity is negligible at this height. The highest electrothermal velocities reported are taken at a height of 20–50 μm above the electrodes [1,13,92]. The lowest electrothermal velocity occurs very close to the electrodes (~5 μm above electrodes in a microchannel with a height of 200 μm) [92]. These observations are in agreement with the physics of the electrothermal effect as it is created in the bulk of the fluid.

Increasing the number of microelectrode pairs has shown to result in increasing ACET fluid flow of high conductivity biofluids at voltages above 4 V_{rms} [92]. In addition, flow velocities measured experimentally are typically smaller than those predicted numerically, sometimes by orders of magnitude, which can be justified to be related to the experimental conditions. Here, we discuss some of the experimental issues reported in the literature.

In their experimental study, Sigurdson et al. showed an electrothermal flow velocity of 100 $\mu m \cdot s^{-1}$, which was reported at a lower voltage in a numerical study [14]. Since their simulations are performed for a 2D geometry, they attributed this discrepancy to neglecting the out of plane heat transfer, which is significant in devices with a small channel height (i.e., ~200 μm). In their later work, since they encountered the same issue (i.e., observing an experimental velocity of 1.5 orders of magnitude lower than the numerical result), they defined an effective voltage [61]. For this matter, a reduction coefficient of 0.38 was introduced to correct the actuation voltage in the numerical simulations. With the modified voltage, the velocity–voltage curve of numerical study closely matches the corresponding curve of the experimental study.

Studying the Joule heating effect, Williams et al. [80] defined a coefficient, E_{rel}, for medium conductivity, to account for the loss in applied electric field, which they attribute to the electrical resistance of their electrode material (i.e., indium tin oxide (ITO)). Therefore, by using a more conductive material for electrodes (e.g., gold), there will be less potential loss (i.e., larger value of E_{rel}). Sin et al. showed a significant deviation of experimentally measured fluid velocity from the theoretically predicted one for fluids with conductivities on the order of 22 $S \cdot m^{-1}$ [60]. The reason for this discrepancy was assumed to be the deterioration of the electrode surface due to electrochemical reactions at high conductivities, which is not accounted for in the numerical model. The same argument is also reported as the reason for deviation of experimental data from simulation in other studies [76,105]. With the surface of the electrode being altered, the electric field, flow patterns, and velocities are no longer predictable as the actual electric potential can be significantly decreased [76]. In addition to high conductivity, increasing voltage above 5.5 V_{rms} is also reported to cause electrode deterioration [76]. Evaporation, and thus a change of the medium's properties, is also mentioned as another possible factor for this discrepancy. In addition, the buoyancy effect can play a role in causing this discrepancy, since a high conductivity medium can experience high temperatures. The ratio of the electrothermal force F_E to buoyancy force F_B can be approximated as the following [4]:

$$\left|\frac{F_E}{F_B}\right| = 7.95 \times 10^{-12}\left(\frac{\nabla T}{\Delta T}\right) \cdot E_{rms}^2 \tag{17}$$

As suggested by this approximation, for buoyancy to be negligible, temperature gradient ∇T needs to be much higher than temperature rise ΔT. As a result, for enhancing the ACET performance, device elements (e.g., electrodes) with relatively high thermal conductivities need to be implemented [4].

6.2. Direction of AC Electrothermal (ACET) Flow

In micropumps with planar electrode configurations, two microvortices on both electrodes in every pair are in competition to determine the direction of net flow [85]. In conventional rectangular asymmetric configurations, since the fluid flow spends more time on the wider electrode, the direction of ACET net flow is determined by the microvortices on the wider electrode [4].

It has been experimentally shown that the direction of flow can be controlled by external heating, i.e., switching the heat flux between wide and narrow electrodes [1,71,106]. This is true for both rectangular symmetric and asymmetric electrode configurations. Additionally, some studies have been performed on the use of unique device configurations to easily toggle the direction of net flow and introduce mixing [113,114], however, these models have not yet been rigorously validated.

6.3. Flow Reversal

One major advantage of ACET is that its velocity is generally independent from frequency, however, at around the crossover frequency, flow reversal occurs, and thus, lower velocities (~10–20% of velocity of low frequencies) are generated [39,69,74,115].

Flow reversal has also been observed in devices with external heating. In 2014, Liu et al. numerically investigated applying a temperature gradient along the channel length on rectangular symmetric electrode arrays and compared it to the conventional asymmetric array [68]. They observed unidirectional pumping in the direction from higher to lower temperatures at 100 kHz (i.e., from the narrow electrode to wide electrode). They also observed that by increasing the frequency to 500 kHz, the direction of flow in a symmetric array reverses but still maintains a unidirectional flow, whereas, in the asymmetric array, the unidirectional flow turns to vortices with no pumping action.

In general, factors other than frequency can also lead to flow reversal. For example, ACEO systems can face flow reversal at higher voltages due to faradaic charging [116]. Transition between ACEO and ACET mechanisms [93] and steric effect can also be important [117,118]. However, none of these reasons can justify the flow reversal in the DC biased AC electrothermal device of Lian et al. [6], which is still open for further investigation.

Flow reversal is considered by most studies as a disadvantage in ACEK due to the uncontrollable and unpredictable nature of this phenomenon. However, some studies have investigated controllable flow reversal by tuning the actuation frequency, switching electric field, and/or applying external heat sources [1,39,68,113,114].

7. Application

Most common applications of the electrothermal effect in micro systems include mixing and pumping of fluids and particle manipulations [1,13,60,62]. A good example of using the electrothermal effect for mixing is immunoassays [11,14,20,61,119–121]. Immunoassays are biochemical tests in which the concentration of macromolecules (e.g., ligands or proteins) in a biofluid is measured by the use of an antibody. The antibody is immobilized on a surface and then the biofluid of interest is introduced above the surface. The concentration of macromolecules is measured by the number of macromolecules attached to the antibodies. The key factor in this process is for macromolecules to contact the antibodies repeatedly so that all possible bindings take place and lead to a precise sensing result. Since this process is diffusion limited, incubation time can take hours [61]. By using the electrothermal effect, the chance of macromolecules reaching the sensing area can be increased, thereby significantly decreasing the response time, enhancing the bindings by seven to nine times within minutes [14,61]. Sigurdson et al. investigated this idea numerically and found that by applying a voltage of 6 V_{rms}, an increase of seven times in the amount of bound antigen can be reached. It should be noted that electrothermal stirring is effective only when Damkohler ≥ 100, i.e., when the process is diffusion limited and not reaction limited. Therefore, electrothermal stirring is not significantly effective for reaction limited systems such as DNA systems. Sigurdson et al. also found that the electrothermal stirring is especially effective in the space above the electrode gap, where the velocity and the concentration gradient are high, making it the optimum place for antibodies to be immobilized [11,14,61]. This was proven experimentally in their later work [61]. Huang et al. found the optimum reaction site to be closer to the negative electrode in a symmetric rectangular pair [122]. In this case, both Damkohler and Peclet numbers need to be considered. Electrothermal stirring is best for mass transport limited regimes, where the Peclet number is relatively low, and convection contributes more to the overall mass transport than diffusion. Therefore, by utilizing the electrothermal effect, the required sample volume can be reduced, leading to higher efficiency devices [14]. It has also been shown that in immunoassay applications, electrothermal force can be a better choice compared to electroosmotic force, as electroosmotic force may cause the antigen–antibody bounds to fall apart [14].

In 2012, Sasaki et al. used a meandering electrode configuration in a Y-shaped channel to mix two high salt content fluids [62]. They reported a fivefold reduction in mixing time compared to diffusional mixing. In their study, the dependence of mixing index on salt concentration, frequency, and mixing time was investigated. It was concluded that the meandering structure was suitable for salt concentrations of 10^{-3} to 10^{-1} mol·dm^{-3}, provided that the frequency lied in the range of 100–200 kHz, which is typical for electrothermal devices. A long range ACET effect, where centimeter scale vortices are generated, can also be used for mixing purposes [70].

Further development on electrothermal based immunoassays was carried out by Liu et al. in 2011 [11]. They decreased the incubation time from 30 min to 3 min by implementing ACET in their immunoassay with the conventional asymmetric electrode array. A study on electrode geometry on capacitive immunoassays showed that electrode geometry is an important component in high electric field configurations with asymmetric geometries rendering higher detection efficiency [123]. Another study suggested that the placement of electrodes on the same surface as the reaction site renders the most efficient configuration for immunoassays [121]. Selmi et al. studied the effect of temperature on immunoassays with asymmetrical electrodes [124]. The use of a pulsed ACET flow and amplitude modulated (AM) sine waves has also been reported for enhancing mixing efficiency in immunoassays [125,126].

Electrothermal effect in conjunction with dielectrophoresis can be used for the trapping and assembly of particles, patterning of colloids, and preconcentration of biological samples for detection and characterization purposes [13,56,59,75,127–131]. Additional studies have shown the enhancement of DNA hybridization with ACET configurations [132]. Recently, the concept of an AC electrothermal micropump was applied to a cell-culture system (with culture media of conductivities up to 2 S·m^{-1}) towards the development of organ-on-a-chip and human-on-a-chip systems [63,133].

Recent studies have also shown that, although other ACEK effects such as DEP can enhance trapping of biological samples, they render nonspecific responses [134].

8. Substrate Material

As electrodes are patterned on a substrate, and the substrate is one of the channel walls through which the heat is dissipated, the choice of substrate material is of great importance [135]. The materials typically used for the fabrication of microfluidic devices include silicon, glass, polydimethylsiloxane (PDMS), and polymethylmethacrylate (PMMA). As a commonly used substrate material, glass has a lower thermal conductivity compared to silicon and can yield higher temperature gradients, and thus, higher rates of electrothermal flow [105]. Heat transfer through the substrate can be controlled by other means too. For example, a thermoelectric cooler under the silicon substrate has been suggested for maintaining the temperature gradient of interest in an ACET device [61]. As opposed to glass, silicon has a higher heat transfer coefficient, and thus, is a good candidate for applications where excessive temperature rise is undesirable [61,85,87,92,136]. Silicon is also admired for its precise geometrical features and low surface roughness [85]. Compared to silicon, the thermal conductivity of PDMS is much lower, causing the temperature rise in PDMS microchannels to be significantly high [92,136].

9. Channel Height

Unlike ACEO, the origin of ACET is the charge density generated in the bulk of the fluid. Therefore, the height of the microfluidic channel plays an important role in the formation of microflows. The microflows will be suppressed when the channel height is too small (<200 μm in a typical microfluidic ACET device) [13,76]. In contrast, ACEO devices utilize a thin layer of electric double layer responsible for dragging the fluid bulk through the microchannel, meaning that reducing the channel height causes the ACEO effect to be more effective.

It has been shown that in two-phase actuation systems, the optimal channel heights are in the range of 500–1000 μm [69]. In a planar configuration, the maximum velocity is reached at the height of ~500 μm, above which no significant difference in the flow rate can be observed [76,85]. In microgrooved structures, however, it is shown that a velocity of five times higher than planar electrode configurations can be reached at a significantly smaller channel height, i.e., ~50 μm, which is 10% of the optimal height in conventional and two-phase systems. This feature of the microgrooved configuration helps further miniaturizing the ACET based devices. As mentioned above, miniaturization is hindered in ACET devices due to dependence of the microflows on the channel height. With increasing the channel height above 50 μm in a microgrooved configuration, the velocity drops but still stays at higher values compared to a planar configuration.

In the case of patterning electrodes both on the top and bottom of the microchannel in an ACET device, increasing the channel height above 200 μm causes the velocity profile to become more similar to that of an ACEO device [104]. A thorough study on the effect of channel height in micropumps can be found in reference [76].

10. Numerical and Experimental Settings

In this section, some key points and influential factors in numerical simulation and experimental setup of ACET devices are briefly reviewed. To assure that the same conditions are valid to apply to other devices of interest, the readers are advised to refer to the articles associated with each statement.

10.1. Numerical Simulation

- Since the wavelength of the electric field is typically larger than the dimensions of the microchannel, electrostatic assumptions can be made [41].
- If electrodes are thin, they do not affect heat transfer [61]. Sufficiently thin electrodes (e.g., ~1200 Å thick) can be assumed isothermal [13,106].
- In typical ACET devices, the ratio of buoyancy force to electrothermal force, i.e., $\left|\frac{F_B}{F_E}\right|$, is estimated to be in the range of $7 \times 10^{-4} - 27 \times 10^{-4}$ [41,85]. When buoyancy is included in the simulations, only a 0.1–0.8% decrease in flow velocity is obtained [41]. Therefore, the buoyancy effect can be neglected in simulations. However, at large length scales and low voltages, it becomes important [41].
- ACEO effect can be neglected at high frequencies and high conductivities [41].
- 2D and 3D simulations usually render the same results. For example, Yuan et al. [41] reported a 1.48% difference in velocities obtained from their 2D and 3D simulations. Therefore, 2D simulations can help saving computational time [41].
- While in most studies electric and thermal fields are considered independent, using the results of Loire et al. [86], Williams [74] conducted numerical simulations with coupled electrical and thermal fields as $\nabla^2 V = \gamma \cdot \nabla V$, where $\gamma = -\beta \nabla T$, instead of the conventional sequential method i.e., $\nabla^2 V = 0$. It was shown that when the temperature rise in a system is >5 K, the two fields can no longer be considered independent [86].
- New methods based on Lattice Boltzman were reported for studying ACET flows [137,138].

10.2. Experimental Setup

- The lighting on the microscope, on which the ACET device is mounted, can play as an external heat source and interfere with the experiments, and thus cause unreliable results. In order to reduce the effects of microscope light, either it needs to be set at its lowest power [13] or a heat absorbing filter between the device and the objective lens needs to be used [62]. Otherwise, illumination effects must be taken into account as an external heat source.
- To reduce the effect of Brownian motion, the average of at least four velocity readings at each voltage setting is recommended to be taken [13].
- If the work involves study of temperature on DEP effect, a non-invasive method (i.e., with no particles involved) must be used to measure temperature in the device. Laser-induced fluorescence (LIF) thermometry, in which a dye is used to measure temperature, is recommended for this purpose [80].
- For measuring velocity, to ensure repeatability, particles must be tracked over at least three pairs of electrodes along the microchannel [41,76].

- For generating effective electric field at the electrode surface, electrodes should be fabricated relatively thin, e.g., 50–100 nm [32].

11. Future Work

Although utilizing the ACET effect for various biomedical applications has been investigated extensively over the past two decades, more investigative work is yet to be carried out to further our knowledge of the phenomenon. Existing drawbacks must be better addressed in order to facilitate utilization of this effect in biomedical laboratory settings. For example, temperature rise, and its detrimental effects on biological samples, is of great concern while using the ACET effect in microfluidic devices. Substrate materials with high heat conductivities, which enable a low temperature rise while keeping high voltages, must be investigated to achieve strong electrothermal flow.

While plenty of novel and efficient electrode designs have been proposed for pumping and mixing applications, the majority of such reports are numerical studies. A lack of experimental studies that would reveal the hidden challenges in applying these novel strategies to real life applications exists in the literature. Such lack of experimental works is mainly due to the limitations in microscale fabrication and electrode degradation occurred at high voltages. To address such fabrication challenges, techniques for low cost fabrication of prototypes of such complicated designs and investigation of different electrode materials and coatings to withstand high voltages [79] are in great demand.

Author Contributions: Writing—original draft preparation, M.N. and A.S.; writing—review and editing, M.N., A.S., T.L., and C.D.; supervision, A.S., C.D.; funding acquisition, C.D.

Funding: This research was funded by a Canadian Natural Sciences and Engineering Research Council (NSERC) Discovery Grant.

Conflicts of Interest: The authors declare no conflict of interest.

Appendix A

The majority of the publications cited throughout the manuscript which have made major contributions are summarized in Table A1.

Table A1. A short summary of relevant articles studying the electrothermal effect, useful for new researchers in this field as a guide to what has been done and those who are working in the field.

Article	Application	Achievement	Specific Observations
[71]	Mixing	Experimental study of illumination-induced electrothermal	The direction of force at high frequencies is form hot regions to cold regions while at low frequencies the opposite is true.
[14,61]	Mixing	Increasing the binding rate and significantly decreasing the incubation time to minutes	Binding rate increased by a factor of nine compared to diffusion-limited reaction
[4]	Pumping	Study of pumping for two electrode configurations of planar asymmetric and orthogonal	Orthogonal configuration yields higher velocities
[13]	Particle manipulation and pumping	Manipulation of particles and fluids of high conductivity at low voltages using a parallel plate and a planar asymmetric electrode configuration	Velocity of 162 $\mu m \cdot s^{-1}$

Table A1. *Cont.*

Article	Application	Achievement	Specific Observations
[93]	Pumping	Numerical and experimental investigation of flow reversal in orthogonal electrodes	Change of flow patterns is a result of change from alternating current electrothermal (ACET) effects to alternating current electroosmosis (ACEO) phenomenon
[6]	Pumping	Applying asymmetry in electric potentials in conjunction with spatial asymmetry	Velocity of 2500 μm·s^{-1}
[62]	Mixing	Introducing meandering electrode configuration with electrothermal effect in a Y-shaped channel	Fivefold reduction of the mixing time of high salt content fluids compared to diffusion-limited methods
[85,97]	Pumping	Introducing microgrooved electrode configuration	Five times increase in pumping rate compared to conventional planar configurations
[69]	Pumping	Introducing two-phase AC signal configuration	25–50% faster flow rates in two-phase configuration compared to the conventional single-phase configuration
[60]	Mixing	Introducing concentric electrode design	Velocity of 70 μm·s^{-1}
[11]	Mixing	Using asymmetric electrodes for immunoassay	Ten times acceleration in binding rate compared to diffusion-limited method (30 min vs. 3 min)
[1]	Pumping	Thermally biased ACET pumping using symmetric and asymmetric electrodes	Velocity of 750 μm·s^{-1}
[75]	Particle manipulation	Using parallel plate (opposing) electrodes in conjunction with thin film resistive heaters	Sorting between 1 μm and 2 μm particles
[92]	Pumping	Study on the effect of the number of electrode pairs over channel length; asymmetric planar electrodes	Increasing the number of electrode pairs helps increase the pumping efficiency
[105]	Pumping	Introducing electrodes both on top and bottom of the microchannel; asymmetric planar electrodes	Opposing electrodes increase the flow rate by 105%
[76]	Pumping	Multiple Array Electrothermal Micropump (MAET) with different actuation patterns and cross sections	Flow rate of 16×10^6 μm^3·s^{-1}
[96]	Pumping	3D circular electrodes	Flow rate of 15×10^6 μm^3·s^{-1}
[104]	Mixing and pumping	Numerical investigation of simultaneous pumping and mixing by introducing microelectrodes on side walls of the microchannel	Mixing efficiency of 80% in <3 min and over a length of <600 μm

Table A1. *Cont.*

Article	Application	Achievement	Specific Observations
[135]	Pumping	Numerical study of multiple array ACET channel	Flow rate of 16×10^6 $\mu m^3 \cdot s^{-1}$
[74,108]	Pumping	Study of using thin film heaters for pumping	2.5 times faster flow rate with thin film heaters compared to Joule heating alone
[63]	Pumping	Application of ACET pumping to cell culture on chip	Flow rate of 44.82 $\mu L \cdot h^{-1}$
[139]	Particle manipulation	Combining ACET and dielectrophoresis (DEP) for detection of circulating cell-free DNA (cfDNA)	Detection of cfDNA in 10 min in concentrations as low as 43 $ng \cdot mL^{-1}$
[140]	Pumping	Numerical and experimental study of the effects of conductivity and channel height on ACET flow	A critical conductivity exists below which there is no net flow and there exists only microvortices
[119]	Mixing	Quantum dot-linked immunodiagnostic assay coupled with ACET mixing	Reduction of detection time from 3.5 h to 30 min using a volume of 2 μL
[59]	Particle manipulation	Development of a mathematical model for rapid electrokinetic patterning (REP) REP based on ACET and DEP	Increasing particle size results in an increase in ratio of ACET to DEP velocity and therefore results in a lower focusing performance
[73]	Mixing	Experimental study of light actuated ACET flow	When AC frequency is above liquid charge relaxation frequency, natural convection is above 35% of the ET flow.
[123]	Mixing	Numerical and experimental comparison of immunoassay performance when using symmetric or asymmetric electrodes	Symmetric and asymmetric geometries render different performance efficiencies only at high electric fields
[102]	Particle manipulation	Numerical and experimental study of electrode material in REP	Titanium electrodes are more efficient than conventionally used indium tin oxide (ITO) electrodes
[141]	Mixing	Numerical and experimental study of AC biased concentric electrodes in biosensors	Faster sensing speed compared to diffusion-limited conditions
[142]	Mixing	Numerical and experimental study of rotating asymmetric electrode pair; Supplying controlled drug concentration to tumor cells	Mixing efficiency 89.12%
[70]	Mixing	Numerical and experimental study of long-range fluid motion induced by ACET microvortices	Centimeter scale ACET vortices are observed
[124]	Mixing	Numerical study of the effect of temperature on binding efficiency in immunoassays	Keeping external surfaces of the microchannel at a constant temperature improves the binding efficiency

Table A1. *Cont.*

Article	Application	Achievement	Specific Observations
[143]	Mixing	Numerical and experimental-3D electrodes embedded inside walls of the channel	Mixing efficiency of 90%
[113]	Pumping	Numerical and experimental study of bi-directional micropump using asymmetric planar electrodes	1500 μm·s^{-1} fluid velocity
[121]	Mixing	Numerical study of electrothermal effect in immunoassays	Placement of electrodes on the same wall as the reaction surface renders the best performance of the biosensor
[126]	Mixing	Study of pulsed ACET flow for detection of dilute samples of small molecules	83% mixing efficiency over a length of 400 μm
[125]	Mixing	Numerical investigation of amplitude modulated (AM) sinewave	100% mixing efficiency with maximum 5.5 K temperature rise
[144]	Mixing	Numerical investigation of the effect of ionic strength on mixing	Mixing efficiency 90%
[134]	Pumping	Experimental study of an immunoassay chip featuring an ACET micropump	Reducing incubation time to 1 min vs. hours in conventional methods
[99]	Simultaneous pumping and mixing	Numerical study of high throughput mixing using opposing asymmetric microgrooved electrodes and symmetric electrode pair	Mixing efficiency of 97.25%
[114]	Simultaneous pumping and mixing	Numerical study of bi-directional pumping and mixing by switching electric potential on planar electrodes	Mixing efficiency of 90% Pumping velocity 90 μm·s^{-1}
[90]	pumping	Numerical investigation of pumping non-Newtonian blood flow	Velocity of 0.02 m·s^{-1}
[89]	Mixing	Numerical investigation of the effect of shear dependent viscosity on mixing efficiency and flow rate using opposing asymmetric microgrooved electrodes and symmetric electrode pair	In similar configurations, dilatant fluids show better mixing efficiency compared to pseudoplastic fluids
[101]	Mixing	Study of arc electrodes in ring-shaped microchamber	100% mixing efficiency at 8 V
[127]	Trapping	Using ACET and DEP to preconcentrate and detect E. Coli	Method can detect concentrations two orders of magnitude smaller than what is possible with diffusion limited methods

Table A1. *Cont.*

Article	Application	Achievement	Specific Observations
[133]	Pumping	Using laser etching on ITO glass to pattern electrodes for pumping cell culture medium in a 3D biomimetic liver lobule model	2 μm·s^{-1} at 5.5 V
[100]	Pumping	Using castellated electrodes; combined DEP and ACET EHD for bioparticle delivery	Negative DEP prevents particles from colliding with channel surfaces; castellated electrodes eliminate ACET vortices
[138]	Pumping	Combining ACET and negative DEP for long range cell transport and suspension in high conductivity medium	DEP is essential for cell suspension under ACET effect
[95]	Simultaneous pumping and mixing	Numerical investigation of 3D asymmetric spiral microelectrode pair	Flow rate 440 μm·s^{-1}
[91]	Pumping	Numerical investigation of the effect of electrode configuration on pumping mechanism of non-Newtonian blood flow	Ring shaped electrodes are the optimal configuration for blood flow pumping
[88]	Pumping, mixing, and trapping	Study of 3D particle-fluid flow under simultaneous effects of ACET, thermal buoyancy (TB), and DEP using multi-layered electrodes	Long range vortices induced by ACET and short-range circulations induced by TB
[77]	Simultaneous pumping and mixing	Introducing two opposing microelectrode arrays placed at an angle relative to channel length	Mixing time reduced by 95% compared to diffusion-limited methods
[72]	Mixing	Study of light induced ACET flow over electrodes of different materials using opposing electrodes	Electrodes with high optical absorption rate and low thermal conductivity are best for effective light-induced heating
[58]	Comprehensive particle and droplet manipulation	Combining ACET and DEP	Particle transit time between multiple branches 0.008 s; droplet sorting purity 90%; particle sorting purity 93%

References

1. Yuan, Q.; Wu, J. Thermally biased AC electrokinetic pumping effect for lab-on-a-chip based delivery of biofluids. *Biomed. Microdevices* **2013**, *15*, 125–133. [CrossRef] [PubMed]
2. Nisar, A.; Afzulpurkar, N.; Mahaisavariya, B.; Tuantranont, A. MEMS-based micropumps in drug delivery and biomedical applications. *Sens. Actuator B Chem.* **2008**, *130*, 917–942. [CrossRef]
3. Zhao, C.; Yang, C. Advances in electrokinetics and their applications in micro/nano fluidics. *Microfluid. Nanofluid.* **2012**, *13*, 179–203. [CrossRef]
4. Wu, J.; Lian, M.; Yang, K. Micropumping of biofluids by alternating current electrothermal effects. *Appl. Phys. Lett.* **2007**, *90*, 234103. [CrossRef]
5. Gagnon, Z.R.; Chang, H.-C. Electrothermal ac electro-osmosis. *Appl. Phys. Lett.* **2009**, *94*, 024101. [CrossRef]
6. Lian, M.; Wu, J. Ultrafast micropumping by biased alternating current electrokinetics. *Appl. Phys. Lett.* **2009**, *94*, 064101. [CrossRef]

Micromachines **2019**, *10*, 762

7. Fuhr, G.; Hagedorn, R.; Muller, T.; Benecke, W.; Wagne, Z. Microfabricated electrohydrodynamic (EHD) pumps for liquids of higher conductivity. *J. Microelectromech. Syst.* **1992**, *1*, 141–146. [CrossRef]

8. Moroney, R.M.; White, R.M.; Howe, R.T. Ultrasonically induced microtransport. In Proceedings of the IEEE Micro Electro Mechanical Systems, Nara, Japan, 30 January 1991; pp. 277–282.

9. Richter, A.; Sandmaier, H. An electrohydrodynamic micropump. In *IEEE MEMS*; IEEE: Napa Valley, CA, USA, 1990; pp. 99–104.

10. Bart, S.F.; Tavrow, L.S.; Mehregany, M.; Lang, J.H. Microfabricated electrohydrodynamic pumps. *Sens. Actuator A Phys.* **1990**, *21*, 193–197. [CrossRef]

11. Liu, X.; Yang, K.; Wadhwa, A.; Eda, S.; Li, S.; Wu, J. Development of an AC electrokinetics-based immunoassay system for on-site serodiagnosis of infectious diseases. *Sens. Actuator A Phys.* **2011**, *171*, 406–413. [CrossRef]

12. Li, D. *Electrokinetics in Microfluidics*; Academic Press: Cambridge, MA, USA, 2004.

13. Lian, M.; Islam, N.; Wu, J. AC electrothermal manipulation of conductive fluids and particles for lab-chip applications. *IET Nanobiotechnol.* **2007**, *1*, 36–43. [CrossRef] [PubMed]

14. Sigurdson, M.; Wang, D.; Meinhart, C.D. Electrothermal stirring for heterogeneous immunoassays. *Lab Chip* **2005**, *5*, 1366–1373. [CrossRef] [PubMed]

15. Ramos, A.; Morgan, H.; Green, N.G.; Castellanos, A. AC electrokinetics: A review of forces in microelectrode structures. *J. Phys. D Appl. Phys.* **1998**, *31*, 2338–2353. [CrossRef]

16. Laser, D.J.; Santiago, J.G. A review of micropumps. *J. Micromech. Microeng.* **2004**, *14*, R35–R64. [CrossRef]

17. Hossan, M.R.; Dutta, D.; Islam, N.; Dutta, P. Review: Electric field driven pumping in microfluidic device. *Electrophoresis* **2018**, *39*, 702–731. [CrossRef] [PubMed]

18. Cao, J.; Cheng, P.; Hong, F. Applications of electrohydrodynamics and Joule heating effects in microfluidic chips: A review. *Sci. China Ser. E* **2009**, *52*, 3477–3490. [CrossRef]

19. Lu, Y.; Liu, T.; Lamanda, A.C.; Sin, M.L.Y.; Gau, V.; Liao, J.C.; Wong, P.K. AC electrokinetics of physiological fluids for biomedical applications. *J. Lab. Autom.* **2015**, *20*, 611–620. [CrossRef] [PubMed]

20. Salari, A.; Thompson, M. Recent advances in AC electrokinetic sample enrichment techniques for biosensor development. *Sens. Actuators B Chem.* **2018**, *255*, 3601–3615. [CrossRef]

21. Ashraf, M.W.; Tayyaba, S.; Afzulpurkar, N. Micro electromechanical systems (MEMS) based microfluidic devices for biomedical applications. *Int. J. Mol. Sci.* **2011**, *12*, 3648–3704. [CrossRef] [PubMed]

22. Xuan, X.; Xu, B.; Li, D. Electroosmotic flow with Joule heating effects. *Lab Chip* **2004**, *4*, 230–236. [CrossRef] [PubMed]

23. Pohl, H.A. *Dielectrophoresis: The Behavior of Neutral Matter in Nonuniform Electric Fields*; Cambridge University Press: Cambridge, UK; New York, NY, USA, 1978.

24. Chakraborty, S. *Microfluidics and Microfabrication*; Springer: Berlin, Germany, 2010.

25. Yang, J.; Huang, Y.; Wang, X.B.; Becker, F.F.; Gascoyne, P.R. Differential analysis of human leukocytes by dielectrophoretic field-flow-fractionation. *Biophys. J.* **2000**, *78*, 2680–2689. [CrossRef]

26. Huang, Y.; Ynag, J.; Wang, X.B.; Frederick, F.B.; Gascoyne, P.R.C. The removal of human breast cancer cells from hematopoietic CD34$^+$ stem cells by dielectrophoretic field-flow-fractionation. *J. Hematother. Stem Cells Res.* **1999**, *8*, 481–490. [CrossRef] [PubMed]

27. Hu, X.; Bessette, P.H.; Qian, J.; Meinhart, C.D.; Daugherty, P.S.; Soh, H.T. Marker-specific sorting of rare cells using dielectrophoresis. *Proc. Natl. Acad. Sci. USA* **2005**, *102*, 15757–15761. [CrossRef] [PubMed]

28. Müller, T.; Fiedler, S.; Schnelle, T.; Ludwig, K.; Jung, H.; Fuhr, G. High frequency electric fields for trapping of viruses. *Biotechnol. Technol.* **1996**, *10*, 221–226. [CrossRef]

29. Morgan, H.; Hughes, M.P.; Green, N.G. Separation of submicron bioparticles by dielectrophoresis. *Biophys. J.* **1999**, *77*, 516–525. [CrossRef]

30. Bhatt, K.H.; Grego, S.; Velev, O.D. An AC electrokinetic technique for collection and concentration of particles and cells on patterned electrodes. *Langmuir* **2005**, *21*, 6603–6612. [CrossRef] [PubMed]

31. Gao, J.; Sin, M.L.Y.; Liu, T.; Gau, V.; Liao, J.C.; Wong, P.K. Hybrid electrokinetic manipulation in high-conductivity media. *Lab Chip* **2011**, *11*, 1770–1775. [CrossRef] [PubMed]

32. Wu, J. Biased AC electro-osmosis for on-chip bioparticle processing. *IEEE Trans. Nanotechnol.* **2006**, *5*, 84–88.

33. Pethig, R. Review Article—Dielectrophoresis: Status of the theory, technology, and applications. *Biomicrofluidics* **2010**, *4*, 039901. [CrossRef]

34. Pethig, R. Review—Where is dielectrophoresis (DEP) going? *J. Electrochem. Soc.* **2017**, *164*, B3049–B3055. [CrossRef]

35. Green, N.; Ramos, A.; Gonzalez, A.; Morgan, H.; Castellanos, A. Fluid flow induced by nonuniform ac electric fields in electrolytes on microelectrodes. I. experimental measurements. *Phys. Rev. E* **2000**, *61*, 4011–4018. [CrossRef] [PubMed]
36. Green, N.G.; Ramos, A.; González, A.; Morgan, H.; Castellanos, A. Fluid flow induced by nonuniform ac electric fields in electrolytes on microelectrodes. III. observation of streamlines and numerical simulation. *Phys. Rev. E* **2002**, *66*, 026305. [CrossRef] [PubMed]
37. Ramos, A.; González, A.; Castellanos, A.; Green, N.; Morgan, H. Pumping of liquids with ac voltages applied to asymmetric pairs of microelectrodes. *Phys. Rev. E* **2003**, *67*, 056302. [CrossRef] [PubMed]
38. Hunter, R.J. *Zeta Potential in Colloid Science: Principles*; Applications; Elsevier Science: Amsterdam, The Netherlands, 2013.
39. Stubbe, M.; Gimsa, J. A short review on AC electro-thermal micropumps based on smeared structural polarizations in the presence of a temperature gradient. *Colloids Surf. A* **2011**, *376*, 97–101. [CrossRef]
40. Studer, V.; Pepin, A.; Chen, Y.; Ajdari, A. An integrated AC electrokinetic pump in a microfluidic loop for fast and tunable flow control. *Analyst* **2004**, *129*, 944–949. [CrossRef] [PubMed]
41. Yuan, Q.; Yang, K.; Wu, J. Optimization of planar interdigitated microelectrode array for biofluid transport by AC electrothermal effect. *Microfluid. Nanofluid.* **2014**, *16*, 167–178. [CrossRef]
42. Wu, J. AC electro-osmotic micropump by asymmetric electrode polarization. *J. Appl. Phys.* **2008**, *103*, 024907. [CrossRef]
43. Bazant, M.Z.; Ben, Y. Theoretical prediction of fast 3D AC electro-osmotic pumps. *Lab Chip* **2006**, *6*, 1455–1461. [CrossRef] [PubMed]
44. Urbanski, J.P.; Thorsen, T.; Levitan, J.A.; Bazant, M.Z. Fast AC electro-osmotic micropumps with nonplanar electrodes. *Appl. Phys. Lett.* **2006**, *89*, 143508. [CrossRef]
45. Ramos, A.; Morgan, H.; Green, N.G.; González, A.; Castellanos, A. Pumping of liquids with traveling-wave electroosmosis. *J. Appl. Phys.* **2005**, *97*, 084906. [CrossRef]
46. Hoettges, K.F.; McDonnell, M.B.; Hughes, M.P. Use of combined dielectrophoretic/electrohydrodynamic forces for biosensor enhancement. *J. Phys. D Appl. Phys.* **2003**, *36*, L101–L104. [CrossRef]
47. Wu, J.; Ben, Y.; Battigelli, D.; Chang, H.C. Long-range AC electroosmotic trapping and detection of bioparticles. *Ind. Eng. Chem. Res.* **2005**, *44*, 2815–2822. [CrossRef]
48. Debesset, S.; Hayden, C.J.; Dalton, C.; Eijkel, J.C.T.; Manz, A. An AC electroosmotic micropump for circular chromatographic applications. *Lab Chip* **2004**, *4*, 396. [CrossRef] [PubMed]
49. Wang, X.; Cheng, C.; Wang, S.; Liu, S. Electroosmotic pumps and their applications in microfluidic systems. *Microfluid. Nanofluid.* **2009**, *6*, 145–162. [CrossRef] [PubMed]
50. Lastochkin, D.; Zhou, R.; Wang, P.; Ben, Y.; Chang, H.-C. Electrokinetic micropump and micromixer design based on ac faradaic polarization. *J. Appl. Phys.* **2004**, *96*, 1730. [CrossRef]
51. Wang, S.C.; Chen, H.P.; Chang, H.C. Ac Electroosmotic Pumping Induced By Noncontact External. Electrodes. *Biomicrofluidics* **2007**, *1*, 034106. [CrossRef] [PubMed]
52. Eckstein, Y.; Yossifon, G.; Seifert, A.; Miloh, T. Nonlinear electrokinetic phenomena around nearly insulated sharp tips in microflows. *J. Colloid Interface Sci.* **2009**, *338*, 243–249. [CrossRef] [PubMed]
53. Wang, S.C.; Lai, Y.W.; Ben, Y.; Chang, H.C. Microfluidic mixing by dc and ac nonlinear electrokinetic vortex flows. *Ind. Eng. Chem. Res.* **2004**, *43*, 2902–2911. [CrossRef]
54. Chang, H.-C.; Yeo, L.Y. *Electrokinetically-Driven Microfluidics*; Nanofluidics; Cambridge University Press: Cambridge, UK, 2009.
55. Morgan, H.; Green, N.G. *AC Electrokinetics: Colloids and Nanoparticles*; Research Studies Press: Boston, MA, USA, 2003.
56. Siva Kumar Gunda, N.; Bhattacharjee, S.; Mitra, S.S.K. Study on the use of dielectrophoresis and electrothermal forces to produce on-chip micromixers and microconcentrators. *Biomicrofluidics* **2012**, *034118*, 1–23. [CrossRef] [PubMed]
57. Chen, D.F.; Du, H. Simulation studies on electrothermal fluid flow induced in a dielectrophoretic microelectrode system. *J. Micromech. Microeng.* **2006**, *16*, 2411–2419. [CrossRef]
58. Sun, H.; Ren, Y.; Hou, L.; Tao, Y.; Liu, W.; Jiang, T.; Jiang, H. Continuous particle trapping, switching, and sorting utilizing a combination of dielectrophoresis and alternating current electrothermal flow. *Anal. Chem.* **2019**, *91*, 5729–5738. [CrossRef] [PubMed]

59. Kim, D.; Shim, J.; Chuang, H.; Kim, K.C. Numerical simulation on the opto-electro-kinetic patterning for rapid concentration of particles in a microchannel. *Biomicrofluidics* **2015**, *9*, 034102. [CrossRef] [PubMed]

60. Sin, M.L.Y.; Gau, V.; Liao, J.C.; Wong, P.K. Electrothermal fluid manipulation of high-conductivity samples for laboratory automation applications. *JALA* **2010**, *15*, 426–432. [CrossRef] [PubMed]

61. Feldman, H.C.; Sigurdson, M.; Meinhart, C.D. AC electrothermal enhancement of heterogeneous assays in microfluidics. *Lab Chip* **2007**, *7*, 1553–1559. [CrossRef] [PubMed]

62. Sasaki, N.; Kitamori, T.; Kim, H.-B. Fluid mixing using AC electrothermal flow on meandering electrodes in a microchannel. *Electrophoresis* **2012**, *33*, 2668–2673. [CrossRef] [PubMed]

63. Lang, Q.; Wu, Y.; Ren, Y.; Tao, Y.; Lei, L.; Jiang, H. AC electrothermal circulatory pumping chip for cell culture. *ACS Appl. Mater. Interfaces* **2015**, *7*, 26792–26801. [CrossRef] [PubMed]

64. Melcher, J.R. Traveling-wave induced electroconvection. *Phys. Fluids* **1966**, *9*, 1548. [CrossRef]

65. Melcher, J.R. Traveling-wave bulk electroconvection induced across a temperature gradient. *Phys. Fluids* **1967**, *10*, 1178. [CrossRef]

66. Fuhr, G.; Schnellet, T.; Wagnert, B. Travelling wave-d riven microfabricated electrohydrodynamic pumps for liquids. *J. Micromech. Microeng.* **1994**, *4*, 217–226. [CrossRef]

67. Castellanos, A.; Ramos, A.; González, A.; Green, N.G.; Morgan, H. Electrohydrodynamics and dielectrophoresis in microsystems: Scaling laws. *J. Phys. D Appl. Phys.* **2003**, *36*, 2584–2597. [CrossRef]

68. Liu, W.; Ren, Y.; Shao, J.; Jiang, H.; Ding, Y. A theoretical and numerical investigation of travelling wave induction microfluidic pumping in a temperature gradient. *J. Phys. D Appl. Phys.* **2014**, *47*, 075501. [CrossRef]

69. Zhang, R.; Dalton, C.; Jullien, G.A. Two-phase AC electrothermal fluidic pumping in a coplanar asymmetric electrode array. *Microfluid. Nanofluid.* **2011**, *10*, 521–529. [CrossRef]

70. Lu, Y.; Ren, Q.; Liu, T.; Leung, S.L.; Gau, V.; Liao, J.C.; Chan, C.L.; Wong, P.K. Long-range electrothermal fluid motion in microfluidic systems. *Int. J. Heat Mass Transf.* **2016**, *98*, 341–349. [CrossRef] [PubMed]

71. Green, N.G.; Ramos, A.; González, A.; Castellanos, A.; Morgan, H. Electric field induced fluid flow on microelectrodes: The effect of illumination. *J. Phys. D Appl. Phys.* **1999**, *33*, L13–L17. [CrossRef]

72. Lee, S.; Kim, J.; Wereley, S.T.; Kwon, J. Light-actuated electrothermal microfluidic flow for micro-mixing. *J. Micromech. Microeng.* **2019**, *29*, 017003. [CrossRef]

73. Kwon, J.S.; Wereley, S.T. Light-actuated electrothermal microfluidic motion: Experimental investigation and physical interpretation. *Microfluid. Nanofluid.* **2015**, *19*, 609–619. [CrossRef]

74. Williams, S.J. Enhanced electrothermal pumping with thin film resistive heaters. *Electrophoresis* **2013**, *34*, 1400–1406. [CrossRef] [PubMed]

75. Velasco, V.; Williams, S.J. Electrokinetic concentration, patterning, and sorting of colloids with thin film heaters. *J. Coll. Interf. Sci.* **2013**, *394*, 598–603. [CrossRef] [PubMed]

76. Salari, A.; Navi, M.; Dalton, C. A novel AC multiple array electrothermal micropump for lab-on-a-chip applications. *Biomicrofluidics* **2015**, *9*, 014113. [CrossRef] [PubMed]

77. Salari, A.; Dalton, C. Simultaneous pumping and mixing of biological fluids in a double-array electrothermal microfluidic device. *Micromachines* **2019**, *10*, 1–11. [CrossRef] [PubMed]

78. Cao, J.; Cheng, P.; Hong, F.J. A numerical study of an electrothermal vortex enhanced micromixer. *Microfluid. Nanofluid.* **2008**, *5*, 13–21. [CrossRef]

79. Lijnse, T.; Cenaiko, S.; Dalton, C. Prevention of electrode degradation in ACET micropumps for biomedical devices. *Alta. BME* **2019**, *1*, 40.

80. Williams, S.J.; Chamarthy, P.; Wereley, S.T. Comparison of experiments and simulation of Joule heating in AC electrokinetic chips. *J. Fluid Eng. T ASME* **2010**, *132*, 021103. [CrossRef]

81. Sridharan, S.; Zhu, J.; Hu, G.; Xuan, X. Joule heating effects on electroosmotic flow in insulator-based dielectrophoresis. *Electrophoresis* **2011**, *32*, 2274–2281. [CrossRef] [PubMed]

82. Hong, F.; Bai, F.; Cheng, P. A parametric study of electrothermal flow inside an AC EWOD droplet. *Int. Commun. Heat Mass Transf.* **2014**, *55*, 63–70. [CrossRef]

83. Green, N.G.; Ramos, A.; González, A.; Castellanos, A.; Morgan, H. Electrothermally induced fluid flow on microelectrodes. *J. Electrost.* **2001**, *53*, 71–87. [CrossRef]

84. Lide, D.R. *CRC Handbook of Chemistry, Physics*, 93rd ed.; Haynes, W.M., Ed.; CRC Press: New York, NY, USA, 2012.

85. Du, E.; Manoochehri, S. Microfluidic pumping optimization in microgrooved channels with AC electrothermal actuations. *Appl. Phys. Lett.* **2010**, *96*, 034102. [CrossRef]

Micromachines **2019**, *10*, 762

86. Loire, S.; Kauffmann, P.; Mezić, I.; Meinhart, C.D. A theoretical and experimental study of AC electrothermal flows. *J. Phys. D Appl. Phys.* **2012**, *45*, 185301. [CrossRef]
87. Hong, F.J.; Bai, F.; Cheng, P. Numerical simulation of AC electrothermal micropump using a fully coupled model. *Microfluid. Nanofluid.* **2012**, *13*, 411–420. [CrossRef]
88. Sato, N.; Yao, J.; Sugawara, M.; Takei, M. Numerical study of particle-fluid flow under AC electrokinetics in electrode-multilayered microfluidic device. *IEEE Trans. Biomed. Eng.* **2019**, *66*, 453–463. [CrossRef] [PubMed]
89. Kunti, G.; Bhattacharya, A.; Chakraborty, S. Analysis of micromixing of non-Newtonian fluids driven by alternating current electrothermal flow. *J. Nonnewton. Fluid Mech.* **2017**, *247*, 123–131. [CrossRef]
90. Ren, Q. Investigation of pumping mechanism for non-Newtonian blood flow with AC electrothermal forces in a microchannel by hybrid boundary element method and immersed boundary-lattice Boltzmann method. *Electrophoresis* **2018**, *39*, 1329–1338. [CrossRef] [PubMed]
91. Ren, Q.; Wang, Y.; Lin, X.; Chan, C.L. AC electrokinetic induced non-Newtonian electrothermal blood flow in 3D microfluidic biosensor with ring electrodes for point-of-care diagnostics AC electrokinetic induced non-Newtonian electrothermal blood flow in 3D micro fl uidic biosensor with ring. *J. Appl. Phys.* **2019**, *126*, 084501.
92. Salari, A.; Navi, M.; Dalton, C. AC electrothermal micropump for biofluidic applications using numerous microelectrode pairs. In *IEEE CEIDP*; IEEE: Des Moines, IA, USA, 2014; pp. 1–4.
93. Yang, K.; Wu, J. Investigation of microflow reversal by ac electrokinetics in orthogonal electrodes for micropump design. *Biomicrofluidics* **2008**, *2*, 024101. [CrossRef] [PubMed]
94. Ajdari, A. Pumping liquids using asymmetric electrode arrays. *Phys. Rev. E* **2000**, *61*, 45–48. [CrossRef] [PubMed]
95. Gao, X.; Li, Y. Biofluid pumping and mixing by an AC electrothermal micropump embedded with a spiral microelectrode pair in a cylindrical microchannel. *Electrophoresis* **2018**, *39*, 3156–3170. [CrossRef] [PubMed]
96. Salari, A.; Dalton, C.A.; Dalton, C. A novel AC electrothermal micropump for biofluid transport using circular interdigitated microelectrode array. In *SPIE BiOS*; Gray, B.L., Becker, H., Eds.; International Society for Optics and Photonics: Bellingham, WA, USA, 2015; Volume 9320, p. 932016.
97. Du, E.; Manoochehri, S. Enhanced AC electrothermal fluidic pumping in microgrooved channels. *J. Appl. Phys.* **2008**, *104*, 064902. [CrossRef]
98. Shojaei, A.; Ramiar, A.; Ghasemi, A.H. Numerical investigation of the effect of the electrodes bed on the electrothermally induced fluid flow velocity inside a microchannel. *Int. J. Mech. Sci.* **2019**, *157*, 415–427. [CrossRef]
99. Kunti, G.; Bhattacharya, A.; Chakraborty, S. Rapid mixing with high-throughput in a semi-active semi-passive micromixer. *Electrophoresis* **2017**, *38*, 1310–1317. [CrossRef] [PubMed]
100. Ren, Q. Bioparticle delivery in physiological conductivity solution using AC electrokinetic micropump with castellated electrodes. *J. Phys. D. Appl. Phys.* **2018**, *51*, aae233. [CrossRef]
101. Meng, J.; Li, S.; Li, J.; Yu, C.; Wei, C.; Dai, S. AC electrothermal mixing for high conductive biofluids by arc-electrodes. *J. Micromech. Microeng.* **2018**, *28*, 065004. [CrossRef]
102. Mishra, A.; Khor, J.; Clayton, K.N.; Williams, S.J.; Pan, X.; Kinzer-ursem, T.; Wereley, S. Optoelectric patterning: Effect of electrode material and thickness on laser-induced AC electrothermal flow. *Electrophoresis* **2016**, *37*, 658–665. [CrossRef] [PubMed]
103. Work, A.H.; Williams, S.J. Characterization of 2D colloids assembled by optically-induced electrohydrodynamics. *Soft Matter* **2015**, *11*, 4266–4272. [CrossRef] [PubMed]
104. Salari, A.; Dalton, C. A novel AC electrothermal micropump consisting of two opposing parallel coplanar asymmetric microelectrode arrays. In Proceedings of the 18th International Conference on Miniaturized Systems for Chemistry and Life Sciences, San Antonio, TX, USA, 26–30 October 2014.
105. Salari, A.; Dalton, C. High efficient biofuid micromixing using ultra-fast AC electrothermal flow. In *SPIE BiOS*; Gray, B.L., Becker, H., Eds.; International Society for Optics and Photonics: Bellingham, WA, USA, 2015; Volume 9320, p. 93201C.
106. González, A.; Ramos, A.; Morgan, H.; Green, N.G.; Castellanos, A. Electrothermal flows generated by alternating and rotating electric fields in microsystems. *J. Fluid Mech.* **2006**, *564*, 415. [CrossRef]
107. Perch-Nielsen, I.R.; Green, N.G.; Wolff, A. Numerical simulation of travelling wave induced electrothermal fluid flow. *J. Phys. D Appl. Phys.* **2004**, *37*, 2323–2330. [CrossRef]
108. Williams, S.J.; Green, N.G. Electrothermal pumping with interdigitated electrodes and resistive heaters. *Electrophoresis* **2015**, *36*, 1681–1689. [CrossRef] [PubMed]

Micromachines **2019**, *10*, 762

109. Wang, Q.; Dingari, N.N.; Buie, C.R. Nonlinear electrokinetic effects in insulator-based dielectrophoretic systems. *Electrophoresis* **2017**, *38*, 2576–2586. [CrossRef] [PubMed]
110. Wang, D.; Sigurdson, M.; Meinhart, C.D. Experimental analysis of particle and fluid motion in ac electrokinetics. *Exp. Fluids* **2005**, *38*, 1–10. [CrossRef]
111. Santiago, J.G.; Wereley, S.T.; Meinhart, C.D.; Beebe, D.J.; Adrian, R.J. A particle image velocimetry system for microfluidics. *Exp. Fluids* **1998**, *25*, 316–319. [CrossRef]
112. Meinhart, C.D.; Wereley, S.T.; Santiago, J.G. PIV measurements of a microchannel flow. *Exp. Fluids* **1999**, *27*, 414–419. [CrossRef]
113. Vafaie, R.H.; Ghavifekr, H.B.; Van Lintel, H.; Brugger, J.; Renaud, P. Bi-directional ACET micropump for on-chip biological applications. *Electrophoresis* **2016**, *37*, 719–726. [CrossRef] [PubMed]
114. Hadjiaghaie, R.; Habib, V.; Ghavifekr, B. Configurable ACET micro—Manipulator for high conductive mediums by using a novel electrode engineering. *Microsyst. Technol.* **2017**, *23*, 1393–1403.
115. Lian, M.; Wu, J. Microfluidic flow reversal at low frequency by AC electrothermal effect. *Microfluid. Nanofluid.* **2009**, *7*, 757–765. [CrossRef]
116. Wu, J.; Ben, Y.; Chang, H.C. Particle detection by electrical impedance spectroscopy with asymmetric-polarization AC electroosmotic trapping. *Microfluid. Nanofluid.* **2005**, *1*, 161–167. [CrossRef]
117. Storey, B.D.; Edwards, L.R.; Kilic, M.S.; Bazant, M.Z. Steric effects on ac electro-osmosis in dilute electrolytes. *Phys. Rev. E* **2008**, *77*, 036317. [CrossRef] [PubMed]
118. Kilic, M.S.; Bazant, M.Z.; Ajdari, A. Steric effects in the dynamics of electrolytes at large applied voltages. I. Double-layer charging. *Phys. Rev. E* **2007**, *75*, 021502. [CrossRef] [PubMed]
119. Yu, C.; Kim, G.-B.; Clark, P.M.; Zubkov, L.; Papazoglou, E.S.; Noh, M.A. microfabricated quantum dot-linked immuno-diagnostic assay (μQLIDA) with an electrohydrodynamic mixing element. *Sens. Actuators B Chem.* **2015**, *209*, 722–728. [CrossRef]
120. Porter, J.M.; Modares, P.; Castiello, F.; Tabrizian, M. Capacitive detection of insulin antibody enhanced by AC electrothermal mixing. In Proceedings of the 2019 IEEE 6th Portuguese Meeting on Bioengineering (ENBENG), Lisbon, Portugal, 22–23 February 2019; pp. 1–4.
121. Selmi, M.; Gazzah, M.H.; Belmabrouk, H. Numerical study of the electrothermal effect on the kinetic reaction of immunoassays for a microfluidic biosensor. *Langmuir* **2016**, *32*, 13305–13312. [CrossRef] [PubMed]
122. Huang, K.-R.; Chang, J.-S.; Chao, S.D.; Wu, K.-C.; Yang, C.-K.; Lai, C.-Y.; Chen, S.-H. Simulation on binding efficiency of immunoassay for a biosensor with applying electrothermal effect. *J. Appl. Phys.* **2008**, *104*, 064702. [CrossRef]
123. Li, S.; Ren, Y.; Cui, H.; Yuan, Q.; Wu, J.; Eda, S.; Jiang, H. Alternating current electrokinetics enhanced in situ capacitive immunoassay. *Electrophoresis* **2015**, *36*, 471–474. [CrossRef] [PubMed]
124. Selmi, M.; Khemiri, R.; Echouchene, F.; Belmabrouk, H. Electrothermal effect on the immunoassay in a microchannel of a biosensor with asymmetrical interdigitated electrodes. *Appl. Eng.* **2016**, *105*, 77–84. [CrossRef]
125. Ghandchi, M.; Hadjiaghaie Vafaie, R. AC electrothermal actuation mechanism for on-chip mixing of high ionic strength fluids. *Microsyst. Technol.* **2017**, *23*, 1495–1507. [CrossRef]
126. Hadjiaghaie Vafaie, R. A high-efficiency micromixing effect by pulsed AC electrothermal flow. *Compel* **2018**, *37*, 418–431. [CrossRef]
127. Frkonja-kuczin, A.; Ray, L.; Zhao, Z.; Konopka, M.C.; Boika, A. Electrokinetic preconcentration and electrochemical detection of Escherichia coli at a microelectrode. *Electrochim. Acta* **2018**, *280*, 191–196. [CrossRef]
128. Kale, A.; Song, L.; Lu, X.; Yu, L.; Hu, G.; Xuan, X. Electrothermal enrichment of submicron particles in an insulator-based dielectrophoretic microdevice. *Electrophoresis* **2018**, *39*, 887–896. [CrossRef] [PubMed]
129. Dies, H.; Raveendran, J.; Escobedo, C.; Docoslis, A. In situ assembly of active surface-enhanced Raman scattering substrates via electric field-guided growth of dendritic nanoparticle structures. *Nanoscale* **2017**, *9*, 7847–7857. [CrossRef] [PubMed]
130. Ramos, A.; García-Sánchez, P.; Morgan, H. AC electrokinetics of conducting microparticles: A review. *Curr. Opin. Coll. Interface Sci.* **2016**, *24*, 79–90. [CrossRef]
131. Zhang, J.; Wang, J.; Wu, J.; Qi, H.; Wang, C.; Fang, X.; Cheng, C.; Yang, W. Rapid detection of ultra-trace nanoparticles based on ACEK enrichment for semiconductor manufacturing quality control. *Microfluid. Nanofluid.* **2019**, *23*, 1–11. [CrossRef]

132. Bottausci, F.; Neumann, T.; Mader, M.A.; Mezic, I.; Jaeger, L.; Tirrell, M. *DNA Hybridization Enhancement in Microarrays Using AC-Electrothermal Flow*; Volume 2: Fora; ASME: Jacksonville, FL, USA, 2008; pp. 629–636.
133. Mi, S.; Li, B.; Yi, X.; Xu, Y.; Du, Z.; Yang, S.; Li, W.; Sun, W. An AC electrothermal self-circulating system with a minimalist process to construct a biomimetic liver lobule model for drug testing. *RSC Adv.* **2018**, *8*, 36987–36998. [CrossRef]
134. Yang, K.; Islam, N.; Eda, S.; Wu, J. Optimization of an AC electrokinetics immunoassay lab - chip for biomedical diagnostics. *Microfluid. Nanofluid.* **2017**, *21*, 1–11. [CrossRef]
135. Salari, A.; Dalton, C. Fluid flow study of an AC electrothermal micropump consisting of multiple arrays of microelectrodes for biofluidic applications. In *SPIE BiOS*; Gray, B.L., Becker, H., Eds.; International Society for Optics and Photonics: Bellingham, DC, USA, 2015; Volume 9320, p. 93200G.
136. Zhang, R.; Jullien, G.A.; Dalton, C. Study on an alternating current electrothermal micropump for microneedle-based fluid delivery systems. *J. Appl. Phys.* **2013**, *114*, 024701. [CrossRef]
137. Ren, Q.; Chan, C.L. Numerical simulation of a 2D electrothermal pump by lattice Boltzmann method on GPU. *Numer. Heat Transf. Part A Appl.* **2016**, *69*, 677–693. [CrossRef]
138. Ren, Q.; Meng, F.; Lik, C. Cell transport and suspension in high conductivity electrothermal flow with negative dielectrophoresis by immersed boundary-lattice Boltzmann method. *Int. J. Heat Mass Transf.* **2019**, *128*, 1229–1244. [CrossRef]
139. Lamanda, A.; Lu, Y.; Gill, N.; Wong, P.K. An electrokinetic microdevice for isolation and quantification of circulating cell-free DNA from physiological samples. In *IEEE Transducers*; IEEE: Anchorage, AK, USA, 2015; pp. 544–547.
140. Tansel, O.; Oksuzoglu, H.; Koklu, A.; Sabuncu, A.C. Electrothermal flow on electrodes arrays at physiological conductivities. *IET Nanobiotechnol.* **2016**, *10*, 54–61.
141. Lee, W.C.; Lee, H.; Lim, J.; Park, Y.J. An effective electrical sensing scheme using AC electrothermal flow on a biosensor platform based on a carbon nanotube network. *Appl. Phys. Lett.* **2016**, *109*, 223701. [CrossRef]
142. Lang, Q.; Ren, Y.; Hobson, D.; Tao, Y.; Hou, L.; Jia, Y.; Hu, Q.; Liu, J.; Zhao, X.; Jiang, H. In-plane microvortices micromixer-based AC electrothermal for testing drug induced death of tumor cells. *Biomicrofluidics* **2016**, *10*, 064102. [CrossRef] [PubMed]
143. Wu, Y.; Ren, Y.; Jiang, H. Enhanced model-based design of a high-throughput three dimensional micromixer driven by alternating-current electrothermal flow. *Electrophoresis* **2017**, *38*, 258–269. [CrossRef] [PubMed]
144. Vafaie, R.H.; Madanpasandi, A. In-situ AC electroosmotic and thermal perturbation effects for wide range of ionic strength. *Aims Biophys.* **2017**, *4*, 451–464. [CrossRef]

MDPI
St. Alban-Anlage 66
4052 Basel
Switzerland
Tel. +41 61 683 77 34
Fax +41 61 302 89 18
www.mdpi.com

Micromachines Editorial Office
E-mail: micromachines@mdpi.com
www.mdpi.com/journal/micromachines